Coping Rituals in Fearful Times

D1740489

Jeltje Gordon-Lennox

Editor

Coping Rituals in Fearful Times

An Unexplored Resource for Healing Trauma

 Springer

Editor
Jeltje Gordon-Lennox
Geneva, Switzerland

ISBN 978-3-030-81536-3   ISBN 978-3-030-81534-9   (eBook)
https://doi.org/10.1007/978-3-030-81534-9

This Springer imprint is published by the registered company Springer Nature Switzerland AG.
The registered company address is: Gewerbestrasse 11, 6330 Cham, Switzerland

Mae-Wan Ho
(1941–2016)
Laureate of the 2014
Prigogine Award

Jacques Ménétrey
(1959–2019)
Advisor for Cultural
Affairs in
Geneva, Switzerland
percussionist and
friend

Graham Stickney
Haber
(1963–2021)
Senior Photographer
at The Morgan
Library & Museum,
New York City, NY,
USA
photo consultant and
family member

Foreword

The idea that rituals are crucial to human behaviour is consistent with the arguments of different social scientists, such as Durkheim, Goffman, Collins, Douglas, Leach, Kertzer, Bell, Alexander, Warner, Shils, and Bellah, who have used this concept to analyse and examine society. However, despite the work of scholars such as these, the concept of ritual has been underutilised, if not often ignored, in sociology and related disciplines.

This is due to the conventional understanding of rituals in sociology and the social/behavioural sciences in general. For instance, it is often assumed that rituals are found only, or far more often, in premodern rather than modern societies. This is due in part to implicit or explicit evolutionary assumptions that depict modern societies as increasingly rational. Rituals are also presumed to be static, unchanging, and fixed in nature. Furthermore, rituals are often thought to occur only or mainly in religious or sacred contexts. And rituals are believed to be of secondary importance to more significant social processes—and epiphenomenal in that they are a product of those processes—which implies they have little effect or significance for people and occurrences in society.

Consistent with these assumptions, we find that many sociologists and others, while attentive to social organisation, pay relatively little attention to culture and/or identity (or personality) structures notwithstanding the contributions of those in social psychology, sociology of emotions, and the sociology (and anthropology) of culture.

For these reasons, rituals are often thought to have limited explanatory value and are often downplayed in social analysis. They remain in various ways invisible to and 'under the radar' of many students of social life and modern society.

In contrast, like this book with its thought-provoking range of approaches to ritual as a resource for healing trauma, structural ritualisation theory (SRT) focuses on the role rituals play in society.[1] Both are grounded on the basic supposition that daily life

---

[1] For a discussion of some of the issues addressed by the SRT perspective, see Knottnerus 2016 (2011) and Knottnerus 1997, 2005, 2009, 2010, 2014a, 2014b, 2015, in press.

is normally characterised by an array of social and personal rituals. Such everyday—often taken for granted—rituals can lead to consequences unanticipated by group members while both being fed by and feeding into larger societal levels of interaction. As such, this theory and the chapters in this collection are directed to rituals that occur in various social settings, e.g. face-to-face interaction, small groups, formal organisations, and society as a whole.

Likewise, just as the presuppositions of SRT support my argument that ritual provides a 'missing link' in sociological thought, they also serve as a general framework for this entire volume on the role of ritual in healing trauma. More precisely:

1. Rituals are found in both premodern and modern societies. Stated somewhat differently, rituals are found all over the world and throughout history. They occur in all societies in one form or another.
2. Rituals occur at and impact micro- and macro-levels of society, ranging from face-to-face interaction and relationships to larger groups and organisations, societies as a whole, and globally. Furthermore, the relations between ritual activities within any particular level and between levels can take many different forms and exhibit various degrees of complexity.
3. Rituals occur in both secular and religious, or more broadly speaking, sacred contexts. They are not restricted to only religious and sacred milieus. Rather, rituals can play a central role in our ordinary, everyday lives and many collective events in the secular realm.
4. Rituals are dynamic in nature and subject to change. They are not always static, fixed, or permanent in nature. While rituals can be enduring they may also be altered for many reasons.
5. Rituals can have consequences for social organisation (i.e. social structure), culture, and identity. These are key dimensions of human and social reality that are of interest to many. Rituals can significantly influence all of these factors.
6. Ritual is a social phenomenon that can be investigated with very different methods and types of evidence, e.g. qualitative and quantitative techniques. Evidence collected by these research strategies can complement, supplement, and validate the findings of different methodologies.
7. The concept of ritual can be linked to perspectives emphasising other social dynamics and issues, i.e. approaches focusing on ritual should be capable of forming linkages or conceptual bridges with other perspectives (what is sometimes referred to as theory integration).
8. The idea of ritual can provide a common vocabulary and framework to study developments occurring in various groups and its individual members. It has analytical value helping us to explain the workings of society.
9. Ritual is a concept that has potential relevance for the multifaceted nature of social life. Consequently, it can be utilised in a wide range of studies given the complexity of human behaviour. In other words, ritual has implications for various dimensions of human reality ranging from personal experiences and individual conduct to diverse kinds of social formations, large and small.

10. Rituals can be of profound importance in social life. They are real and consequential for humans, albeit in numerous and varied ways.
11. Rituals have great explanatory value. Simply stated, they help us understand different aspects of social behaviour in a multitude of situations.

In my work, SRT directly addresses these issues and concerns because it, amongst other things, uses the idea of ritual to explain different social developments. It provides formulations that focus on ritual dynamics taking place in many settings throughout the world and in different historical periods, e.g. small groups, corporations such as Enron, ethnic communities in urban areas or small towns, slave societies in the antebellum south, health-care facilities, especially nursing homes, the emergence of golf in the USA, youth groups in nineteenth-century French elite schools, the socialisation of youth in ancient Sparta, and political systems ranging from the Nazi party to the Khmer Rouge. For all these reasons, ritual and perspectives using this concept provide a missing link in sociological analysis and more generally the social/behavioural sciences.

So too the work presented in this book is consistent with and builds upon these assumptions. The following chapters demonstrate in different ways how ritual provides a missing link in our understanding of human behaviour and social dynamics. More accurately, these chapters in both bold and more implicit, subtle ways demonstrate how critically important trauma is in human lives and the importance of ritual for coping with and mitigating the deleterious effects of traumatic experiences.

I would also emphasise that this volume's focus on ritual and trauma parallels and contributes to one line of research in SRT dealing with the disruption of ritualised practices. While I have not focused on the concept of 'trauma', it is directly relevant for much of this research and significantly expands the scope of what others and I have studied and thought about. Our work focuses on disruptions, deritualisation, and reritualisation, i.e. breakdowns of social and personal rituals, their consequences, and the ways people may cope with such experiences by reconstituting old or new ritualised activities.

More precisely, disruption refers to events or conditions that interrupt the rituals that people normally engage in. Deritualisation involves the breakdown or loss of previously engaged-upon rituals, i.e. the cessation of ritualised practices that occur in our daily, taken-for-granted lives. This can be an extremely difficult condition for individuals and groups; it can be confusing, uncomfortable, disorienting, aversive, and destabilising. Reritualisation refers to the re-creation of rituals after disruption and deritualisation. The re-creation of rituals (and patterns of such practices) helps people achieve, amongst other things, a sense of direction, a meaningful focus, coherence in their perceptions and behaviour, stability in their lives, and a sense of security. Thus, the re-enactment of rituals serves as a buffer to disruptive occurrences. They enable people to cope with problematic situations such as these.

Research has examined these ideas in a variety of settings, including internment in concentration camps during the mid-twentieth century, the displacement of youth during China's Cultural Revolution, disasters in general, the impact of earthquakes on a major city in Nepal, dark ages/periods of ecological degradation in ancient China, task groups in a laboratory experiment, the Khmer Rouge in Cambodia with particular attention to ritual and social control, rituals engaged in by victims of disasters, i.e. tornadoes striking two American cities, and the ritual and social dynamics of crews on polar expeditions from the mid-nineteenth century to the mid-twentieth century.

Because of the contributions to this new book edited by Jeltje Gordon-Lennox, I have gained a much better appreciation for how most if not nearly all of the cases studied involve different forms of trauma.

Indeed, the chapters in this book (and the research just mentioned) offer many ideas and raise many questions about the nature of rituals and how rituals may help people cope with the disruptions caused by, for instance, (a) hurricanes, earthquakes, and tornadoes, (b) actions often coercive in nature such as wars, colonisation, terrorism, internment, or imprisonment, (c) long-term, unsafe, stressful, and isolated ventures and settings such as expeditions, space missions, pandemic lockdowns, or refugee camps, and (d) social, economic, scientific/technological, and political developments in the modern world and to a certain degree in premodern societies. All of these situations and others not mentioned can involve trauma and the potential for different kinds of ritualised responses.

Overall, the goals of this volume are clearly stated by the editor. They are to examine a wide variety of approaches that are potentially relevant to the issue of ritual as an asset for responding to trauma and to focus on what it means to ritualise in ways not impeded by mistaken presumptions and ways of thinking. In doing so, it suggests new points of view for examining an extremely wide range of ritual practices. And it would promote further examination of the ways and reasons ritual takes such varied forms in different epochs, regions of the world, cultures, and even within specific societies.

I will not address in detail the organisation of the book and the authors and contents of the different chapters because Jeltje Gordon-Lennox does so in exemplary fashion in the "Introduction". I will say that the contributors' focus reflects the just described objectives of the editor and many of the suppositions that underlie the belief that ritual provides a missing link for understanding human behaviour and the workings of society. My comments will, therefore, be of a much more limited and general nature concerning the assumptions and goals of the authors.

Certainly, the contributions to this volume rest on the assumption that rituals occur in the present and the past, i.e. modern and premodern times, and in all types of societies and regions of the world. For instance, some chapters focus on the archaeological study of mortuary rituals in prehistory, healing rituals in ancient China (481–221 BCE), and rituals such as processions, festivals, and the wearing of masks in medieval Europe. Other chapters, on the other hand, concentrate on the modern world, sometimes in a very broad manner, and other times giving more attention to modern Western nations. And certain chapters focus on rituals in South Sudan (both today and in the past), and contemporary Afghanistan, Ukraine (and the Soviet Union), Nepal, Brazil, and Switzerland.

While the focus of the contributions to this book is on a more micro-level since they are examining trauma in the lives of people and their enactment of rituals, they also often give attention to more macro-levels of society including the ways the latter may influence the former. For instance, we learn how widespread events and practices such as wars and political conflict in a country, the treatment of patients in hospitals (large organisations) and medical professions (which operate at a national level), and hazardous environmental practices, some of which are influenced by large-scale political, economic, and corporate entities, influence the lives of individuals, the suffering they experience, and the rituals they may turn to.

So too we find that while some rituals are of a more sacred nature, others occur in very secular settings. Contrast, for example, the more traditional, ancestral rituals of the Dinka in South Sudan, rituals which brought a sacred quality to the secular environments of a hospital, those of farmers struggling to survive in Nepal or Brazil, or the routine use of the Internet by persons around the world. In the latter case, we learn, however, that sometimes when people mourn and memorialise online the death of certain persons this type of collective activity takes on, albeit in a temporary manner, a special meaning for all concerned, a quality that might be considered to be sacred in nature.

Actually, this online collective event also shows how rituals are dynamic. For instance, a new online ritual has emerged in recent years in which individuals respond to the death of a person in a manner that is quite different from traditional practices. And another example is the recent development of a community-based ritual whose goal is to facilitate healing and reconciliation in war-torn Afghanistan.

Given the potency and ubiquitous nature of many rituals, it should not be surprising that they have consequences for social organisation, culture, and identity (and personality). For example, it is argued that responses to social dislocation in the modern world can lead to different forms of powerful addictions, which may be reduced if not eliminated through the cultivation of particular kinds of communal rituals. When this happens the social worlds people live in and the structure of their relations to other groups can radically change along with the culturally shared beliefs and sentiments they share with others and the way they see themselves, i.e. their identity. Moreover, in a study of people's responses to terror, we find that when individuals in two quite different groups engaged in special rituals, their sense of despair or trauma was ameliorated and the values presumably shared by group members were affirmed.

Of course, we should remember that the study of ritual is not limited to any particular technique. Many different methodologies can be employed to study this phenomenon as the chapters in this volume clearly demonstrate. Some of the methodologies used include interviews, field observations of specific groups and/or collective events, personal (autobiographical) accounts, and the examination of historical evidence such as ancient texts or objects, e.g. masks. So too some studies are based on the examination of other research employing similar or different methods whether they be quantitative or qualitative in nature.

Moreover, some contributions are informed by or directly build upon different conceptual frameworks and arguments, i.e. a bridge, linkage, or integration of theoretical ideas. For instance, one chapter presents an extremely sophisticated formulation, polyvagal theory, which primarily deals with neurophysiological processes involving the brain and a major nerve system. Yet, being receptive to observations and insights concerning ritual, the theory also addresses behaviours that are found in ancient rituals and their relevance for people's physiological condition in contemporary society. In a quite different vein, the chapter on addiction in modern societies, which takes a psychological—or perhaps I should say psychosocial—approach, draws upon the ideas of such diverse thinkers and scholars as Plato, Karl Polanyi, and Émile Durkheim.

The issues and investigations discussed so far clearly show how ritual is a concept that provides a common vocabulary and set of ideas for studying and better understanding different groups and persons. The scope of the concept is to be sure wide-ranging. As one example of how this concept can be used, consider the investigation of the Chernobyl nuclear disaster and the doctors and scientists who at great risk documented the health costs of the contamination for countless persons and who developed protocols for curing and preventing some of the medical problems many people experience. Their dedication to such endeavours was grounded in the rituals of science which they were committed to, ranging from the collection of evidence and developing explanations for what they found to the need to disseminate this information to others and provide care for all those who were sick. The value of ritual is quite apparent in this chapter on Chernobyl and especially the discussion of it by Jeltje Gordon-Lennox in the introductory chapter.

The cases examined in this volume also demonstrate the value of ritual for understanding the different dimensions of human reality. It is a concept that can be used in many kinds of studies whether they address the impact of the scientific and medical professions to which one belongs, the ways ritual can help people cope with disturbing experiences, or other investigations.

Finally, all the chapters in this volume and the other studies referred to here show how profoundly important ritual is in the lives of humans, past and present. And how important it is for dealing with harmful and emotionally disturbing disruptions in our lives, perhaps especially in the world we live in today. For all of these reasons, ritual has great explanatory value.

To summarise, this book (and SRT) rests on the fundamental assumption that ritual is a key dimension of social behaviour as are other aspects of social conduct such as rationality, symbolic interpretation, or emotions. Put somewhat differently, *ritual is like an engine that drives much of social life*, sometimes quite intensely (Knottnerus, 2016 [2011]). This driving force, particularly as it is exposed in these chapters, remains largely unacknowledged at a time when ritual greatly influences how society handles the trauma of a pandemic.

In closing, the chapters in this volume are extremely timely and relevant, highly engaging, thought-provoking, sometimes quite moving, and simply put exceedingly interesting if not fascinating to read. It is a very valuable work, not only for all the reasons previously discussed, but because it paves the way for more theorising, research, reflections, and conjectures about ritual, trauma, and their effects on individuals' psychological states, the psychosocial reality of group members, and broader collective phenomena within society. I urge all concerned with such issues to read this book.

J. David Knottnerus
Emeritus Regents Professor of Sociology,
Oklahoma State University,
Stillwater, OK, USA
e-mail: david.knottnerus@okstate.edu

References

Knottnerus, J. D. (1997). The theory of structural ritualization. In B. Markovsky, M.J. Lovaglia, & L. Troyer (Eds.), *Advances in group processes* (Vol. 14, pp. 257–279). Greenwich, CT: JAI Press.

Knottnerus, J. D. (2005). The need for theory and the value of cooperation: Disruption and deritualization. *Sociological Spectrum, 25*(1), 5–19.

Knottnerus, J. D. (2010). Collective events, rituals, and emotions. In S.R. Thye & E.J. Lawler (Eds.), *Advances in group processes* (Vol. 27, pp. 39–61). Bingley, UK: Emerald Group Publishing Limited.

Knottnerus, J. D. (2014a). Emotions, pride and the dynamics of collective ritual events. In G. B. Sullivan (Ed.), *Understanding collective pride and group identity: New Directions in emotion theory, research and practice* (pp. 43–54). London and New York: Routledge.

Knottnerus, J. D. (2014b). Religion, ritual, and collective emotion. In C. von Scheve & M. Salmela (Eds.), *Collective emotions: Perspectives from psychology, philosophy, and sociology* (pp. 312–325). Oxford: Oxford University Press.

Knottnerus, J. D. (2015). Structural ritualization theory: Application and change. In J. D. Knottnerus & B. S. Phillips (Eds.), *Bureaucratic culture and escalating world problems* (pp. 70–84). London and New York: Routledge.

Knottnerus, J. D. (2016 [2011]). *Ritual as a missing link: Sociology, structural ritualization theory and research.* London and New York: Routledge.

Knottnerus, J. D. (in press). *Polar expeditions: Rituals, crews, and hazardous ventures.* London and New York: Routledge.

Acknowledgements

This collection grew out of a professional curiosity about how people ritualise to deal with feelings of helplessness in the face of danger and uncertainty. In August 2017, nearly half of the chapters of this volume had been completed when a worksite accident radically altered my circumstances. I fell seriously ill; life as I knew it came to an abrupt halt as the subject of my project became a personal challenge: my family and I needed to move from immobilised shock to action in order to seek appropriate medical care and restore a sense of safety to our home (see the final chapter).

I am grateful to our family doctor, Françoise Chuat Vuissoz, for down-to-earth practical advice about clean-up, to Carmen who made it happen, and to Alexey V. Nesterenko at BELRAD for guidance about cures with concentrated apple pectin. This simple product, used by populations in areas of the Republic of Belarus to lower radiation contamination from Chernobyl, helped our bodies deal with exposure to high levels of lead (Pb). David R. Chettle at McMaster University was amazing; we are grateful to him for the X-ray fluorescence analysis of our bones—as well as for his patience and wisdom in addressing our many questions. My heartfelt thanks go to Stephen W. Porges for suggestions on how to cope with profound fatigue and loss of concentration and to neurologist Roman Sztajzel for identifying the neurological issues. Nadja Holfeld (physical therapist), Petra Zahn (osteopath), and Olivier Gavin (ergotherapist) helped me face the day-to-day trials of neurological dysfunction.

During those dark months, somatic experiencing (SE) professors Liana Netto and Sônia Gomes, as well as rolfer Annika Sundell, taught me how to use my body as a resource. My SE group was generous with moral support, in particular Christina Heinl, Maria Lucia Reenberg, Anita Ogilvie, Daya Cabestany, Zena Takieddine, and Nelia Reyes. Special thanks go to Isabel Russo, Tara Rice, Leslie Jagoe, and Jussi Pellonpää for reading parts of the manuscript and making insightful comments. I am most indebted to Matthieu Smyth for his collaboration, guidance, contributions and unfailing belief in the pertinence of the project. The creativity of my colleagues at the European Ritual Network inspires this book.

My warmest thanks go to the team at Springer Nature for taking up the challenge of publishing this new collection on coping rituals and the healing of trauma. In particular, I am grateful to Shinjini Chatterjee my commissioning editor for her enthusiastic support, as well as to my project coordinator Shanthi Ramamoorthy and project managers Marianathan Sandou and Ilakya Raghuraj for their equanimity and professionalism.

Expression of my profound appreciation and praise goes to all of the contributors for their patience and trust. I am touched by the humility, mutual respect, and dedication of the authors, especially those who have collaborated on chapters. Mae-Wan Ho's remarkable work features in this volume thanks to Peter Saunders and Eva Sirinathsinghi of Science in Society. Thanks too to Adam Petrusek for his contribution. The dedication and efficiency of editor Alexandra Holmes and indexer Margaret de Boer gave the book its professional polish. Graham Stickney Haber, our photo consultant, died as this book was going to print. Graham was a deeply compassionate person, and unfailingly generous and helpful when we came to him for help or advice of any kind. Last but certainly not least, I am grateful to Ian for sharing my life over these last 30 years. His steady presence transformed our recent challenges into another few bumps along our anything but dull path together.

Contents

Part III   Global Threat, Trauma, and Ritual

About the Editor

Jeltje Gordon-Lennox, MDiv, is a psychotherapist trained in body-based approaches and world religions. Her research and practice is influenced by her life experiences in conflict zones on several continents, in particular her work with the International Committee of the Red Cross. She has written five practical guides on secular ritualising, two in French and three in English. This collection continues the conversation on ritual and trauma started in *Emerging Ritual in Secular Societies: A Transdisciplinary Conversation* (2017, Jessica Kingsley Publishers). Jeltje lives with her husband and their two children in Switzerland. *Website:* gordon-lennox.ch *E-mail:* Jeltje@Gordon-Lennox.ch

List of Figures and Tables

Chapter – 'Dead Land Dead Water'

Introduction

Ritual, Dignity, and the Fragility of Life

Jeltje Gordon-Lennox

> Tunisian Chamseddine Marzoug is more famous for his 'cemetery of the forgotten' than he is for his prowess as a fisherman. Several times a week he digs graves to give a certain dignity to the unidentified migrants whose bodies wash up on the beaches of Zarzis. When he is not burying their remains, he marks and lays flowers on their tombs.[1]
>
> Death row inmate Roger McGowen declared his innocence from the start. The Afro-American man who has been imprisoned since 1986 undertook, alone, a long spiritual journey that led him from rage at the system and victimhood to compassion, forgiveness, freedom, and unconditional love.[2]
>
> Many of the islands in the world, including those off the coast of the Netherlands, may be lost to the sea over the next 20–30 years. Through the

(continued)

---

[1] Chamseddine's cousin Captain Bourassine and his colleagues rescued hundreds of drowning migrants as they fished the Mediterranean Sea. Italian authorities punished the Captain by imprisoning him, confiscating his fishing boat, and dumping his catch back into the sea (Harari & Maillard, 2019).

[2] For over 20 years, Swiss writer and sociologist Pierre Pradervand has been in contact with Roger McGowen. This support turned McGowen into a spiritual model for hundreds around the world. See Pradervand's book, *Messages of Life from Death Row* (2009).

J. Gordon-Lennox (✉)
Psychotherapist ASP, Geneva, Switzerland
e-mail: jeltje@gordon-lennox.ch

© The Author(s), under exclusive license to Springer Nature Switzerland AG 2022
J. Gordon-Lennox (ed.), *Coping Rituals in Fearful Times*,
https://doi.org/10.1007/978-3-030-81534-9_1

fragile beauty of the spectacular migration of monarch butterflies, Dutch artist Desirée Dolron communicates urgent and foreboding concern for the effects of environmental degradation on insect and human movements.[3]

'I've experienced grief. But until the COVID-19 pandemic, I had not felt a part of an immense and widespread global grief,' writes Zenobia Jeffries Warfield.[4] 'Now we're in it too—right alongside people in Italy, Japan, South Africa, and most other countries. In my hometown of Detroit, the number of new cases and deaths climbs. While African Americans are only 14% of the state's population, we're 33% of the COVID-19 cases and 41% of the deaths. Amid so much grief and renewed anger at the inequalities and injustices, we're witnessing a rising up of our communities on a breathtaking scale. We have seen the power of community … to change the future … to draw on reserves of resilience … to look out for all people … to cultivate joy despite fear.' (Warfield, 2020)

As unpredictability escalates on a global scale, the aptitude shown by people like Chamseddine, Roger, Desirée, and Zenobia to apply their creativity and imaginative faculties to coping with fear and despair through ritualising becomes ever more crucial for human survival. Humankind has always used creativity, imagination, and ritualising to deal with not only practical needs like food, clothing, and shelter, but also to cope with threats such as illness, transience, and death. Nearly a century ago, anthropologist Bronisław Malinowski recognised the human psycho-physiological need for ready-made or invented ritual acts to palliate anxiety and fear in uncertain times[5] (Malinowski, 1948 [1925]).

The chapters in this volume build on previous conversations of scholars on ritualising, such as those initiated by Malinowski, and the 1966 symposium on *Ritualization in Man*, which was broadened in *Secular Ritual* (1977) and in *Emerging Ritual in Secular Societies* (2017).[6] While the symposium directs our attention to

---

[3] In addition to 'Monarch' (2018), Desirée Dolron has created other installations with similar concerns such as 'Complex Systems' (2017), 'I will show you fear in a handful of dust I' (2016), 'Uncertain TX' (2016), and 'Xteriors' (2001–2005). See website: Dolron (2021).

[4] Zenobia Jeffries Warfield, an Executive Editor at YES! Magazine, is a mother and journalist who is always searching for ways to better herself, her family, her community, and the world.

[5] When man realises his impotence, 'his anxiety, his fears and hopes, induce a tension… His nervous system and his whole organism drive him to some substitute activity. The man lost at night in the woods or the jungle, beset by fear, [may act] like an animal which attempts to save itself by feigning death. These reactions are natural responses to such a situation, based on a universal psycho-physiological mechanism. They engender what could be called extended expressions of emotion in act and in word.' Malinowski refers to these 'expressions' as 'spontaneous ritual and verbiage' (1948 [1925], pp. 60–62).

[6] The symposium was organised by Julian Huxley. Women picked up the concern for non-religious ritualising: Sally Falk Moore and Barbara G. Myerhoff edited *Secular Ritual* (1977) and I edited *Emerging Ritual in Secular Societies* (2017).

how ritualisation—a highly condensed verbal and behavioural language—serves to perpetuate and communicate essential knowledge, the second conference attempts to define ritual and unyoke it from religion. The third book points to the role and function of non-religious ritualising in contemporary multicultural societies. The contributors to this fourth volume, *Coping Rituals in Fearful Times*, hone in on how the transformative power of ritualising can mitigate trauma by sustaining wellbeing and connectedness in an era of uncertainty, where human interactions are increasingly virtual.

It is important to clarify at the outset what this book is and what it is not. The volume does not pretend to be an anthropological or scientific discussion of what ritual is or of ritual efficacy, be it secular or sacred. It does not intend to pit traditional rites against emerging ritual, nor does it stand within or promote a particular theoretical perspective on ritual. Rather, the goal of this work as a whole is to highlight what it means today to ritualise at the systemic edge by drawing attention to the immense spectrum of human ritual activity that brings to bear on the issue of ritual as a resource for treating trauma.

Doing Ritual

Our prehistoric ancestors left us few clues as to how they made and performed their rituals. Yet the objects they left behind attest to their great imagination and creativity in ritualising the bittersweet events of life. As noted, ritual practice is inherently tied to fearful situations because it appears to reduce the anxiety that impedes normal functioning by replacing it with a feeling of control (Malinowski, 1948 [1925]; Knottnerus, 1997, 2016; Sosis & Handwerker, 2011; Snodgrass et al., 2017; Lang et al., 2020).

Like our ancestors, we too create and 'do ritual' from birth to death to meet our biological needs for safety, deal with strong emotion, and be with others. Today, we tend to align ourselves with a number of loose 'tribes'—physical or virtual—that share our interests, concerns, ideals, and fears, such as our yoga class, online gamer group, or the parents with whom we wait and chat each day at the gate of our children's school. We commonly gravitate towards collective and personal rituals that help us be and feel like a balanced, good, and typical person in our corner of the world.

In and of itself, ritualmaking is neutral; it's what we humans do. Yet just as ritual practice can promote love, healing, and social cohesion, its shadow side enforces a peculiar model of order that favours the injustice, inequality, and fragmentation that foments war, hatred, fake news, and racism. Imposed rituals—such as those enforced by legal systems, psychiatric hospitals, religious institutions or sects, or the workplace—are readily perceived as manipulative.

Fig. 1 Spectrum of ritual activity. Ritual represents a vast array of human activity that spans from elaborate public ceremony to intimate personal habits. This hexagonal isometric representation of the spectrum is nearly an infinity symbol. Rather than divide ritual activities into two spheres, the relatively few rigid lines create a mirror effect. Ceremonial ritual—long the main focus of ritual study—represents only a small part of the public/formal sphere where it is juxtaposed against superstition. Likewise, consumerism is juxtaposed against obsessive-compulsive behaviours. Common gestures and words, like shaking hands and saying 'hello', also figure in the model.
| © J. Gordon-Lennox

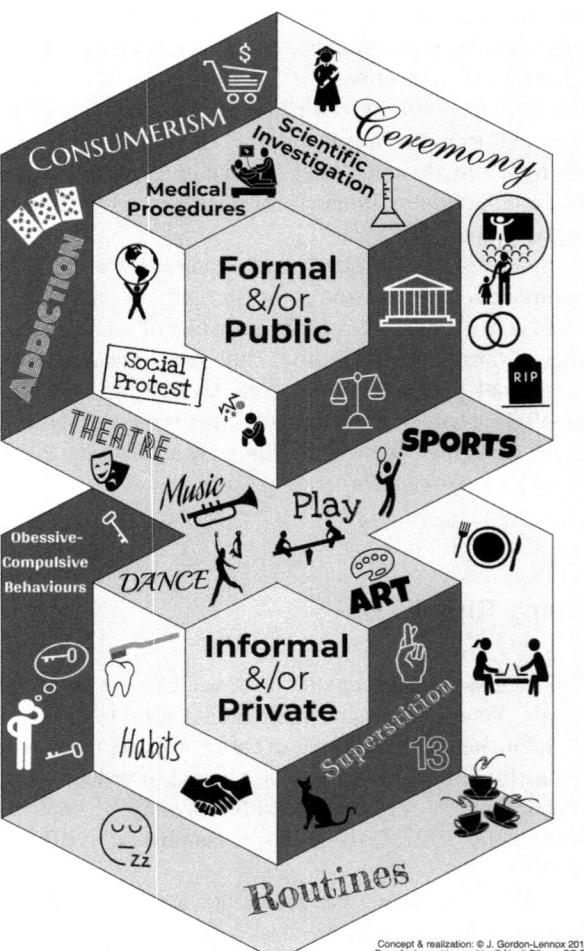

Concept & realization: © J. Gordon-Lennox 2018
Base for isometric graphic: © Nevit Dilmen CC BY

Spectrum of Ritual Activity

A strikingly vast array of activity is associated with ritual, from elaborate public ceremony to intimate personal habits. Jury trials, execution of criminals, and scientific symposia are marked by ritual tradition, repetition, and invariance, whereas a family meal often features spontaneity, irregularity, and variety (see Fig. 1). Ritual studies scholar Catherine Bell's model of ritual practice, built on the work of Bourdieu (1977), recognizes a continuum of structuring and structured practices, ranging from everyday routines to highly ritualised ones with no clear separation of 'sacred' and 'profane' (2009 [1992]).

What becomes evident from this perspective is the natural shift in focus away from the rituals people 'have' towards *how* and *why* people 'do ritual'. Viewing ritual as a spectrum of activity highlights the importance of considering vantage point. From fiancés' point of view, their wedding is about love; paramount to the state is the change in the couple's civil and tax status. Most people bury their mother only once; scholars look at the big picture to see ritual repetition. Variation, not uniformity, in what, when, where, how, and how frequently rituals are created and practised is the norm across all times and peoples. The concept of *ritual acculturation*[7] fits with our notion of how and why ritual practice changes from one era, region, and culture to the next. Ritual is, and has always been, culturally constructed. Perhaps this is why ritual has managed for so long to elude, if not outrightly defy, definition.

Going Global

With lucid foresight, anthropologist Jack R. Goody noted in the 1970s that the term 'ritual' was being defined and applied so widely among scholars, and others, that it was becoming a useless 'global construct' that could mean everything and anything; he appealed for a 'revitalising paradigm shift'[8] (1977, pp. 34–35). Ritual as a global construct first materialised with the post-war Beat culture of the 1950s and the hippie movements of the 1960s. These countercultures gave rise to creative ritual performances where 'improvisation, direct experience, immediacy, and spontaneity were priorities' (Aukeman, 2016, p. 107). Since then, traditional rites appear to cohabit in an uneasy ménage à trois with new rituals on one side and ritual theories on the other.[9]

At the heart of this uneasiness is a diehard notion of ritual as religious, if not sacred, with inherent transformative power that sets it apart from daily life, protects it from scrutiny, and welds it to prescribed order, formalism, repetition, invariance, and traditionalism.[10] This narrow view contributes to the very aura that has recently

---

[7] Social psychologist Batje Mesquita's work on acculturation led to a concept she calls *emotion acculturation* (Mesquita et al., 2016). Adapted to the notion of ritual, the term *ritual acculturation* shows how one set of *ritual and ritual practices* are replaced by other sets to affect meaning, relationships, and health.

[8] Goody's article appears in *Secular Ritual*. It refers to scientist Thomas Kuhn's popularised concept of 'paradigm shift', which argues that scientific advancement is not evolutionary but a 'series of peaceful interludes punctuated by intellectually violent revolutions', during which 'one conceptual worldview is replaced by another' (Kuhn, 1996 [1962], p. 10).

[9] Bell questions whether it is even possible to generate a theoretical model that simultaneously respects the ritual, the participants, and the social scientists analysing them (1993). Indeed, it is difficult to measure the implications of analysing 'vantage point', that is, what traditional or new rituals feel like for those at the centre of the ritual, participants, observers, and scholars without influencing either the process or the outcome.

[10] Certain archaeologists decry the confusion created by the identification of some practices and occasions as ritual and others as domestic (Brück, 1999; Bradley, 2003). The 'term ritual or rather

made ritual so popular—and so marketable. With a savvy makeover that untethers ritual from religion—while retaining a transcendent aura—marketing researchers and advertisers have successfully turned ritual into a global consumer construct and set in motion a revitalised paradigm shift quite different from the one imagined by Goody. The reconditioning of ritual as a consumer construct forces us to take a deeper look at the context in which ritual went global and what might distinguish it from non-ritual.

Global Dislocation[11]

Ritual was, in fact, just one of many constructs co-opted to serve global consumer mechanisms. Sociologist Saskia Sassen tags the 1980s as a watershed for how wealth and power are accumulated and redistributed. The innovative global mechanisms invented to replace primitive accumulation of wealth (e.g. peasant and communal land ownership) and boost consumerism range from the logistics of outsourcing to the algorithms of finance. These same mechanisms cause extensive habitat destruction that expels both the monarch butterflies portrayed by Desirée Dolron's art and the people who wash up and are buried on a Tunisian beach by Chamseddine Marzoug.

> [T]hese types of development [have led to greater inequality,] shrinking economies in much of the world, escalating destructions of the biosphere all over the globe, and the reemergence of extreme forms of poverty and brutalization where we thought they had been eliminated or were on their way out. Today, the structures through which concentration happens are complex assemblages of multiple elements, rather than the fiefdoms of a few robber barons. (Sassen, 2014, p. 12)
>
> [The] vast destructive processes that produce dislocations are by-products of the pursuit of what powerful actors are after. The fact that people are pushed out of having reasonable lives can be [considered] a secondary effect. (Sassen, 2015, p. 178)

For perhaps the first time in human history, the usual discourse on poverty, racism, and injustice does not suffice to grasp the sources, impact, and sheer scale of the new socioeconomic and environmental dislocations. Granted, Zenobia Jeffries Warfield's description of the 2020 lockdown in Detroit leaves no doubt that the poor and those living in unjust contexts remain the most vulnerable. Yet the subtle shift to global dislocation is even more brutal than the actual targeting of the poor and defenceless. It causes suffering and fragmentation across the board, touching public arenas (political, economic, social, cultural, spiritual) as well as our intimacy (emotional, sensory, sexual, neuronal, relational). Moreover, the underlying

---

the more useful concept of ritualization as a process (cf. Bell, 2007, 2009) has the potential to inform our understanding of situations and phenomena which are definitely not religious in any sense' (Hamilakis, 2011, p. 211).

[11] In chemistry, the term 'dislocation' refers to irregularities in the fine structure lattice of an otherwise normal crystal. In this context, the word refers to subtle, and not so subtle, irregularities in the fine structure lattice of modern society.

brutalities that lead to dislocation are too often produced and reproduced by the complexity of the knowledge and technology that we have come to depend on—and may even admire.

Expulsions at the Systemic Edge

Sassen qualifies these dislocations as a type of expulsion (2014, 2015) that affects not only the small farmer in India, Brazil, or Switzerland but also the upper-middle-class family with a subprime mortgage and the hospital doctor forced to choose between 'risk management' and the Hippocratic Oath.

People living at what Sassen calls the ambiguous or systemic edge[12] are forced to migrate when they are expelled from their professional livelihood or driven from their homes. Like Roger McGowen, they are expelled from, and kept from going back to, their usual living spaces when full occupancy—which ensures good returns on shareholder investments in privatised prisons, refugee camps, and other such facilities—prevails over rehabilitation of the condemned or returning the displaced to normal life. Along with monarch butterflies, migrants, prisoners, and others are lost in transit and become invisible. Expelled men, women, and children do not count at all; they do not even figure into the picture (2015).

The sheer complexity of these global assemblages makes it difficult, if not impossible, to trace responsibility for the expulsions produced. The ruthless logic of the systems' assemblages makes it equally hard for those who reap the benefits to feel responsible for any 'collateral damage' caused to people or the environment. Where no responsibility can be attributed or assumed, the risks of fragmentation and dislocation are increased a hundredfold, along with fear, uncertainty, stress, and trauma.

Transforming Trauma

When our sense of control over our environment is seriously compromised, fear of not being able to protect ourselves and those we love puts us at risk of psycho-physiological trauma, and thus of physical and mental illnesses, and social isolation.

---

[12] Sassen conceives of the systemic or ambiguous edge as the point where a condition takes on a format so extreme that it cannot be easily apprehended. This ambiguous edge signals the existence of conceptually subterranean trends. The proliferation of these edges means that each major domain has its own distinctive systemic edge (2015).

Fear Is a Vital Emotion

Fear is a deep primitive emotional and physical reaction to threat that can save but also kill. Just like the pain that pulls our hand from a hot stove, fear can pull us out of harm's way. A quick peek at top selling books like *Feel the Fear and Do It Anyway* (Jeffers, 1987) reveals that the fight-flight reaction to threat is well known. Action can indeed jump-start us out of fear and lead to power and agency. Paradoxically, life itself may be threatened when fear initiates metabolic shutdown (Scaer, 2014 [2001]; Schore, 2002).[13]

Being 'scared to death' refers to the rabbit-in-the-headlights reaction. Society often blames victims of aggression who collapse into submission rather than fleeing or fighting back. Feeling trapped by a danger that cannot be escaped activates a primitive immobilisation reaction that can result in death feigning or even death. A strong tie between fear and imagination can enhance or inhibit these processes. Political powers, for example, are very much aware of the value of fear—and the illusion of safety—for social control. In the 1950s, the threat of nuclear war coalesced with individual and communal senses of insecurity to make fear a global phenomenon (Bash, 2014). The phenomenon amplified in the 1980s with a systemic logic that gave rise to dislocation from and destruction of the very biosphere that makes life possible.

Embodied terror in the face of real or imagined inescapable threat of death can become a reoccurring experience, even a permanent existential state, that has profound impact on our capacity for love and work. Terror, along with the isolation that is at the core of intense fear and chronic trauma, literally reshapes both brain and body (Van der Kolk, 2014). Sadly, any action that feels like re-exposure to threat may well increase the risk of trauma or re-traumatisation (Levine, 2015, 2018).

Transforming Tragedy

How is it that some people appear to bounce back and others remain mired in inertia? 'Despite the seemingly boundless human predilection to inflict suffering and trauma on others, we are also capable of surviving, adapting to, and eventually transforming traumatic experiences' observes psychologist Peter A. Levine (2015, p. ix). 'In the case of biologically important concerns,' affirms ethologist Ellen Dissanayake,

> people *do something more* to try to influence or ensure the outcome they desire. They make things associated with these matters special—extraordinary—even to the point of creating complex physical and mental constructions or ways of doing things that are not obviously

---

[13] There is evidence that early relational trauma is particularly expressed in right hemispheric deficits in the processing of social-emotional and bodily information. Maltreated and neglected children diagnosed with PTSD manifest right lateralised metabolic limbic abnormalities. As adults they may regress to an infantile state when confronted with severe stress (Schore, 2002).

relevant to the vital matter at hand. These complex 'constructions' or 'ways' are called rituals or ceremonies. . . . existential uncertainty—leading to emotional investment or 'caring about'—was the original motivating impetus for the invention of ritual in humans. One can observe in every society that rituals are meant to affect biologically important states of affairs whose attainment is uncertain. (Dissanayake, 2017, p. 92, emphasis added)

Art-filled ritual practices (actions and words) fulfil a basic human need. Catherine Bell saw ritual as 'a culturally strategic way of acting in the world' (Jonte-Pace, 2009, p. vii). She used the concepts of ritualisation and the process of embodiment to distinguish ritual—with its privileged, significant and powerful aspects—from non-ritual acts. Bell also positioned ritual as something that creates or generates—rather than just expresses or reflects—meaning and structure (Nilsson Stutz, 2014).

Against all odds, we humans can intuitively but consciously use our imagination to bridle fear and creatively transform utter helplessness and incomprehensible tragedy. Zenobia, Roger, Desirée, and Chamseddine feel compelled to ritualise their fear of ongoing threat through embodied words and acts. Ritualisation gives their concerns structure and meaning. It palliates anxiety and cultivates hope and joy. One day at a time. Resilience, the term often used to describe the amazing human capacity for coping with hardship, requires a sense of safety that goes far beyond the objective removal of threat.[14] Safety is the *state of feeling safe.*

Searching for Safety

In times of global dislocation, with the precipitous advent of what French philosopher Frédéric Lenoir dubs the 'ultramodern era',[15] the power of formalised religious and civic rites has waned in contemporary society. Rarely has humankind had to face such rapid large-scale change with so few meaningful traditional rituals.[16] As our vertical and horizontal safety zones swiftly disappear, people experience dislocation and fragmentation as never before. 'We have killed the gods,' remarks Lenoir, 'we have abolished or erased our borders. It is within ourselves that we must now find these "safety zones"' (Lenoir, 2012, p. 64).

---

[14] Sociologist David Knottnerus points to evidence from studies of concentration camps and other forms of internment to show how important personal and group rituals can be for enabling people to cope with highly disruptive experiences. Whether the disruption is positive (e.g., marriage, beginning a new job, birth of a baby) or negative (e.g., death of close friend or relative, loss of a job, divorce, internment), those who are able to create new or reconstitute old ritualised practices in the midst of such disruption are best able to cope with, adapt, maintain relationships with others, and survive their immediate situation (Knottnerus, 2016).

[15] Lenoir prefers the term 'ultramodern' to 'postmodern' because the latter gives the false impression that we are disenchanted with the myth of progress and the modern process, which is contradicted by the unprecedented acceleration of modernity (critical reason, individualisation, globalisation).

[16] 'There is really very little evidence to suggest that ritual in general declines per se. It may be more accurate to say that it shifts', observed Catherine Bell (2009 [1992], p. 166).

Like Lenoir, Peter A. Levine encourages the building of 'islands of safety' within ourselves to keep from being overwhelmed by the aftereffects of highly charged life experiences. He observes that 'whether we are survivors of trauma or simply casualties of Western culture' (2010, p. 256), we may suffer from 'an impairing disconnection from [our] inner sensate compass' (2010, p. 355). This is experienced as fragmentation or disembodiment. Inordinate amounts of energy are needed just to keep these sensations under control—usually at the expense of concentration, the ability to memorise, and the ability to pay attention to what is happening around us. Neurologist Robert Scaer calls this frightening experience an aberration of memory[17] (Scaer, 2014 [2001]). The inability to live fully in the present impedes adequate preparation for the future, which in turn wreaks havoc on health and social relationships such as marriages, families, and friendships (Scaer, 2005, 2012).

A Science of Safety

Researcher Stephen Porges' complex polyvagal theory (2011) is a 'science of safety' that can help us learn how to deal with the challenges of daily life like threat and stress but also how to enjoy happiness; it involves 'feeling safe enough to fall in love with life and take the risks of living' (Dana, 2018, p. xvii). According to Porges' polyvagal theory, the autonomic nervous system responds hierarchically with evolutionarily newer circuits inhibiting older circuits (see Porges chapter). The newest pathway, the myelinated ventral vagal pathway (social engagement and connection), enhances co-regulation of our nervous system through contact with people, pets, and other mammals. Co-regulation is identified by Porges as a biological imperative. In dangerous and life-threatening contexts, the autonomic nervous system shows adaptive flexibility through two more primitive neural pathways which regulate defensive responses to threat: the sympathetic nervous system (mobilisation) and the dorsal vagus (immobilisation).

Hailed as a missing link both by clinicians and individuals who have experienced trauma, polyvagal theory (2011) provides plausible neurophysiological explanations for what trauma feels like from the inside. It puts words on often inexplicable sensations with new vocabulary such as *neuroception*, which differentiates 'detection without awareness' from our normal perception with awareness. Neuroception describes how the autonomic nervous system informs us on very subtle levels about our state of safety and any threat of danger, be it within our bodies, in our surroundings, or in our relationships with others. Therefore neuroception does not

---

[17] In the case of trauma, memory imprints (known as 'engrams') are experienced, not as a recurring recollection of a terrible event that happened in the past, but as overwhelming life-threatening physical sensations in the immediate present (Levine, 2015). These physical sensations are ever the more frightening in that they may be tied to events that we do not remember and then triggered without warning by anything—a sudden noise, a smell, a taste, a colour, or a tone of voice—usually totally unattached to a conscious memory of an event.

tell us about *what* we are or *who* we are but *how* we are; it involves a *deep sense* of safety or threat that can be influenced by caring or traumatic experiences (see Fig. 2).

Ritual as Neural 'Exercise'

Just as polyvagal theory constitutes a missing link to understanding trauma, it may also be a missing link to apprehending ritual practice. Especially since, as Levine (2017, n.p.) observes, 'ritual has been an overlooked asset to the healing of trauma and to restoring broken connections'. Stephen Porges affirms that 'a careful investigation of many rituals results in the discovery that the rituals are functional exercises of vagal pathways' (cf. Porges' chapter in Part I). This discovery indicates that ritual practice, whether composed of secular, routine, or religious elements (see Fig. 2), has less impact over *what* or *who* we are than over *how* we are. In other words, ritual process is more about wellbeing and personal subjective feelings of connectedness at the most basic levels than about meaningfulness or structures.

The function of ritual practice 'may be different from that of the narratives upon which religions were based' notes Porges. 'The narratives are attempts to fulfil the human need to create meaning out of uncertainty and to understand the unknowable mysteries of the human experience in a dynamically changing and challenging world' (cf. Porges' chapter in Part I). Yet, as ritual practice exercises the vagal circuit and neuroception becomes perception with awareness, it may open the way to create or generate meaning and structure, as suggested by Bell (2009 [1992]).

Ritual must feel right to be right (Holloway, 2015). This right feeling rests on what I call *embodied intentionality*, which differs from mindfulness[18] in that it involves 'gut-up' simultaneous dual awareness of the ritual experience and of one's own bodily sensations. The embodiment of intentionality informs the inexplicable, mystical, or even magical sense of connectedness inherent in ritualising. My experience of accompanying traumatised people in therapy and through ritual practice confirms that ritualising is not therapy but it can indeed be therapeutic. The psycho-physiological manifestations of embodied intentionality in ritual—as felt by the practitioner and observed by therapists or celebrants—resemble the release of tension observed during trauma resolution.

An imbalance in ritual practice that favours awareness of experience over sensations fosters mental constructs (living in one's head) that can traumatise or re-traumatise through unbidden memories and feelings of isolation, shame, guilt, or disgust. A mourner who is medicated 'to get through the funeral', for example, may feel unsupported in grief. Conversely, the mourner, or trauma victim, who

---

[18] A recent study reports that 82% of people with negative experiences during mindfulness meditation had experienced fear, anxiety, or paranoia (Lindahl et al., 2017). Trauma survivors may be more at risk for the overwhelm and dissociation that leads to this kind of negative experience.

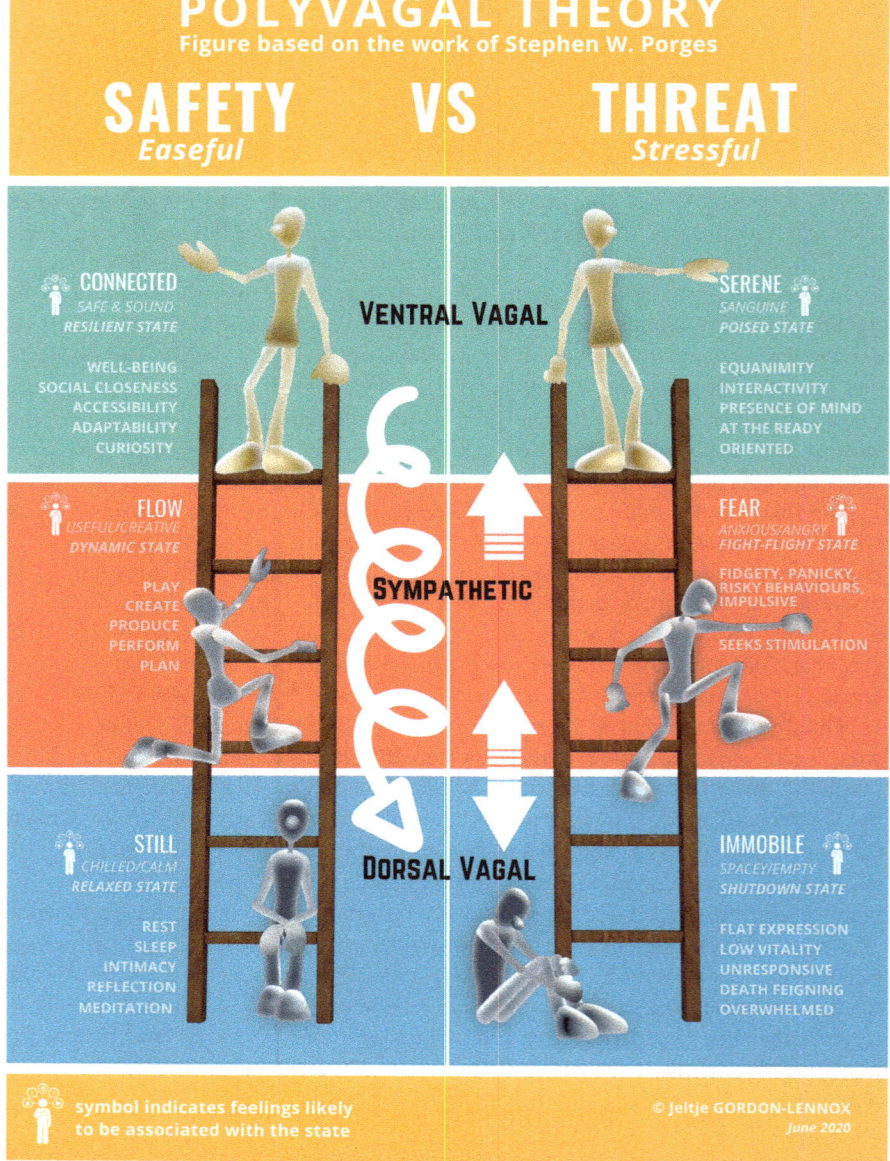

Fig. 2 Safety and threat. Whether we feel safe or threatened, transitioning between the three states (vagal, sympathetic, or dorsal) follows the same course on the diagram. The curly downward arrow (centre left) suggests a certain fluidity of movement between connection (ventral vagal) and *either* flow or fear (sympathetic) *or* calm and terror (dorsal vagal). The thicker arrows (centre right) indicate that moving in the opposite direction, from a dorsal vagal to a ventral vagal state, requires passing through the sympathetic state. Keeping this in mind helps us understand why we may stretch upon awaking from sleep as well as why the likelihood of risky behaviours, or even suicide, may increase as a person under threat moves from intense overwhelm (dorsal state) to action (sympathetic state). | © J. Gordon-Lennox

tracks bodily sensations without simultaneous attention to the experiential ritual context/content may find temporary relief for anxiety, grief, or terror but retain a sense of incompleteness. This embodied process of experience with sensation, in a context that feels safe, is what enhances self- and co-regulation.[19]

I posit that embodied intentionality may also be what distinguishes as ritual or non-ritual the practice of activities like habits, routines, obsessive-compulsive behaviours, and addiction.

Deb Dana's Four Rs

Therapist Deb Dana's four Rs polyvagal approach to trauma survival (Dana, 2018, p. 7) can be applied to ritual practice:

- 'Recognise the autonomic state.'

  Look for and reinforce cues of safety during ritualising. Feeling safe is essential to an inner sense of control over out-of-control situations.
- 'Respect the adaptive survival response.'

  Compassionate accompaniment from a ventral vagal state facilitates changes in the physiological state of the trauma survivor that supports feelings of safety and allows connection through ritual.
- 'Regulate or co-regulate into a ventral vagal state.'

  A person whose default mode is the sympathetic state may be able to self-regulate during ritualising by imagining mastering fear and anxiety. Or they may co-regulate via the ventral vagal state of a compassionate person. Those immobilised in a dorsal vagal state need an extra step to move towards the ventral vagal state. The presence of a compassionate person can encourage *remembering how it once felt to feel safe*, thus fostering a gentle shift to the sympathetic state and then to connectedness in the ventral vagal state.
- 'Re-story.'

  From a place of safety, the individual can perform voluntary behaviours during rituals that involve breathing, postural shifts, and vocalisations that functionally exercise the vagal circuit—without fear or overwhelm—to promote, reinforce, and strengthen states of calmness in the ventral vagal state.

---

[19] Experience with sensation, in a context that feels safe, would appear to determine the effectiveness of trauma treatments that involve repetitive gestures such as David Berceli's Trauma/Stress Releasing Exercises (TRE), Francine Shapiro's EMDR, and Gary Craig's Emotional Freedom Technique (EFT).

Ritualising at the Systemic Edge

The challenges of ritualising in fearful times—particularly during ongoing threat—are not the same for people who belong to mainstream spiritual systems, or for the excluded and the expelled. Extreme conditions constitute a systemic edge for ritualisation that alerts us to more moderate, hidden trends relating to the spiritual dimension of contemporary human experience. Lenoir's concern about the absence of vertical and horizontal safety zones alludes to subterranean spiritual trends—trends that are invisible to conventional ways of seeing, being, and elucidating meaning. Clues to their presence lie in seemingly negligible contradictions in values taught and practised, or in minor disagreements between leaders and practitioners about ritual function and traditional narrative. These subterranean trends originate in the decline of Western-style global religions and capitalisms, the escalation of environmental destruction, and the rise of complex forms of knowledge and technology (Sassen, 2014, 2015).

None of the extreme conditions exposed by Desirée, Zenobia, Chamseddine, and Roger are easily captured by the standard measures of governments and experts. Each condition signals a different major domain with its own distinctive systemic edge. Together, as these conditions draw attention to the expanding populations of migrants, prisoners, the expelled or the ignored, they point to the proliferation of many such ambiguous edges across the world. These four intentionally embodied ritualisations manage to transform unbearable helplessness and trauma at the ambiguous edge into something extraordinary and healing for both the ritualmaker and for those who bear witness to them.

Through ritualmaking we mark time, seasons, and space; we anchor our memories and ourselves in the present, rather than in the past or the future. As our rituals are regularly updated, they provide guidance to individuals and groups on how to act in the current environment. The amazing capacity of our human imagination to intentionally produce and embody abstract signs and sounds is essential to the creative processes of healing trauma and ritualmaking. These processes result in what is perhaps one of the most astonishing aspects of ritual practice and trauma treatment: that is, feeling how the subtle interplay between individual and collective imagination allows human beings to enhance both self- and co-regulation (Tateo, 2016), thus preserving their humanity and dignity.

Chapter Summaries

The scholars, artists, and practitioners who wrote these chapters specialise in fields as diverse as anthropology, bioarchaeology, digital culture, ancient Chinese studies, psychiatry, political science, bioenergetics, ritual anthropology, philosophy, and psychotherapy, as well as in the political and social role of museums and peace studies. Although the contributors may present divergent views on ritual and trauma, taken together these chapters demonstrate how embodied intentionality in ritualising can initiate healing processes and mitigate the negative impact of trauma on individuals, collective groups, and even global systems.

Part I: Trauma and Ritual in Other Times and Places

Bioarchaeologists Liv Nilsson Stutz and Aaron Jonas Stutz open this volume with their unique perspective on the material traces of ritual in prehistory. Rituals have long been part of the deeply human strategy for coping with death. While signs of recourse to ritual as a tool for healing and comfort upon the loss of a loved one appear often, archaeological records also reveal evidence of how rituals have been used to respond to catastrophic events or weaponised to increase fear or power.

Stephen W. Porges, researcher in traumatic stress and originator of the landmark work on polyvagal theory, presents a model that shows how specific voluntary behaviours (e.g. breathing, vocalisations, and posture shifts), which characterise ancient rituals and form the core of contemplative practices, can trigger a physiological state that fosters health and optimises subjective experiences. Porges' model emphasises that, in order to experience the positive benefits of these practices, the physical context must be perceived by the practitioner as calm and soothing, and above all, safe.

Researcher Ori Tavor examines key passages from classical Chinese texts written during the Warring States period (481–221 BCE), one of the most tumultuous and traumatic periods in Chinese history. Drawing on contemporary scientific work on how music impacts emotion regulation, Tavor offers valuable insights on the transformative effects on individuals and groups of ancient ritual events that combined music, dance, and the offering of sacrifices.

Ritual anthropologist Matthieu Smyth closes the section on ancient rituals with his chapter on the festivals that developed in medieval urban culture to counter the brutal hierarchical social systems of that era. Smyth observes that the transformative power of the old animistic rituals to conjure and regulate human reactions to threat was carried over to the new rituals that arose. He raises the question: What might we learn from these ancient European rituals about how the resolution of fear and anxiety in the face of death and calamity is transformed and adapted from one age to the next?

Part II: The Role of Ritual in Healing Trauma

The fact that medical interventions can cause trauma is a well-kept secret. Patients, their entourage, and medical professionals may all find themselves at the sharp end of medical care. Three therapists, Robin Karr-Morse, Juan Carlos Garaizabal, and Jeltje Gordon-Lennox, collaborated to produce this chapter. They know from experience that the rituals surrounding medical procedures can serve to reduce fear, prevent disruption, and maintain order, or cause medically induced trauma. Regardless of the outcome of the medical intervention, rituals based on transparency and compassion can effect healing among all concerned at all levels of medical care.

The following two chapters examine the impact of ritualising in the context of armed conflict.

Kenyan philosopher Alex N. Kamwaria presents his unique perspective in a sensitive investigation of the use of ancestral rituals and Western psychosocial interventions to heal trauma and facilitate reconciliation within the Dinka community in South Sudan. While Western interventions view trauma as 'post', Dinka victims experience their suffering as being very much in the present. Kamwaria's findings support the implementation, wherever possible, of cultural alternatives to the new international mechanisms such as truth commissions, criminal courts, and tribunals.

The Memory Box Project is a unique community-led effort that addresses traumatic experiences stemming from decades of political violence in Afghanistan. Memorialisation specialist Sophia Milosevic Bijleveld shows how ritualising memory can address impunity and provide victims and their community with a symbolic sense of justice—even in the absence of traditional rituals and political will to implement transitional justice mechanisms—by giving voice to victims and ensuring that the crimes and violence in Afghanistan are not forgotten. Early August 2021, as this volume goes to print, the Afghan government has collapsed and the Taliban rule the streets of the capital. Milosevic Bijleveld writes from Kabul: 'I am 24/7 on trying to put Afghan Victims Archives to safety and evacuating the staff...'.

Interdisciplinary researcher Sasha A.Q. Scott closes this section with his work on online rituals of solidarity in the wake of highly public death events. Scott argues that social media memorialising cultivates a 'sense of ritual'. The making and sharing of memorial videos, posting selfies of solidarity, a strategic use of hashtags, the remediation of symbolic imagery, and—perhaps most importantly—the collectives that form online around these digital expressions all serve an important role in the healing process. Spontaneous and unstructured online rituals circumvent ritual specialists and collapse barriers of time and space, accelerating and amplifying the scale of public mourning in unscripted, highly creative, and personalised ways.

The recent spate of funerals by zoom during successive pandemic lockdowns is further evidence of how bans on physical presence are creatively skirted in order to meet our fundamental human need for ritual and social connection. Without a doubt, Scott's arguments regarding the power of online ritual responses to public death events also apply to private digital memorialising.

Part III: Global Threat, Trauma, and Ritual

Psychologist Bruce K. Alexander collaborates with anthropologist Matthieu Smyth to present a radical rethink of the nature and healing of addiction, which, in recent years, has gone global. Its range extends far beyond drugs and alcohol to gambling, shopping, romantic love, video games, religious zealotry, television viewing, internet surfing, and emaciated body shapes. Plato explicitly invites us to face the nightmare possibility that, in a dislocated, deteriorating society, the most addicted people may become the political leaders. The ferocity of their addiction to power can make others turn to them in the vain hope of finding safety and a secure identity.

Peacebuilder Lisa Schirch analyses political leadership and what it means to ritualise in an age of terror and violent extremism. Schirch defines and compares the ritualistic aspects of terrorism and extremism and shows how different groups and people neutralise the effects of these violent acts through ritual. She documents and describes ritual responses, such as 'magical resistance' and 'binding spells', used to counter the perceived violent extremism evident in the 2016 election and mandate of Donald Trump, 45th president of the United States.

On January 6, 2021, just months after Schirch's article was submitted for publication, Trump supporters violently stormed the United States Capitol where a joint session of Congress was beginning the Electoral College vote count to formalise Joe Biden's victory. In response, 22-year-old poet Amanda Gorman stirred hope and awe at the 46th president's inauguration with her poem 'The Hill We Climb': 'We've braved the belly of the beast. We've learned that quiet isn't always peace, and the norms and notions of what "just" is isn't always justice' (Gorman, 2021). Eight months later, like Milosevic Bijleveld, Schirch is working desparately to protect her Afghan staff from Taliban reprisals. The magical resistance cited in Schirch's chapter, Gorman's words and Schirch's actions channel feelings of fear and powerlessness in an attempt to restore a sense of connection and community.

The chapter on nuclear threat was woven together from Mae-Wan Ho's work by Alexy V. Nesterenko, Odile Gordon-Lennox, and Peter Saunders. It explores how a group of doctors and scientists risked their lives and careers to help people living in the areas most contaminated by the Chernobyl fallout. Using the rituals of scientific method, they discovered a simple treatment based on apple pectin that can clear radionuclides from the body. In the aftermath of Fukushima, it was found that certain seaweeds have similar properties. These solutions offer hope for future generations of Chernobyl and Fukushima victims.

The collection closes with three cases: a flash flood, the effects of pesticide exposure, and a worksite accident—all of which affect my family, directly or indirectly. Although these kinds of accidents occur regularly on large and small scales around the world, few studies consider the risk of psychosocial trauma resulting from the ongoing risk of such environmental hazards. In this chapter I examine how, as chronic threat pushes ordinary people out of the common to the systemic edge, ritualising may serve as an adaptive mechanism for coping with trauma in the midst of unremitting threat.

References

Aukeman, A. (2016). *Welcome to Painterland: Bruce Conner and the Rat Bastard Protective Association*. University of California Press.

Bash, L. (2014). The globalisation of fear and the construction of the intercultural imagination. *Intercultural Education, 25*(2), 77–84.

Bell, C. (1993). The authority of ritual experts. *Studia Liturgica, 23*(1), 98–120.

Bell, C. (Ed.). (2007). *Teaching ritual*. Oxford University Press.

Bell, C. (2009 [1992]). *Ritual theory, ritual practice*. Oxford University Press.

Bourdieu, P. (1977). *Outline of a theory of practice* (Vol. 16). Cambridge University Press.

Bradley, R. (2003). A life less ordinary: The ritualization of the domestic sphere in later prehistoric Europe. *Cambridge Archaeological Journal, 13*(1), 5–23.

Brück, J. (1999). Ritual and rationality: Some problems of interpretation in European archaeology. *European Journal of Archaeology, 2*(3), 313–344.

Dana, D. A. (2018). *The polyvagal theory in therapy: engaging the rhythm of regulation (Norton Series on Interpersonal Neurobiology)*. WW Norton & Company.

Dissanayake, E. (2017). Ethology, interpersonal neurobiology, and play: insights into the evolutionary origin of the arts. *American Journal of Play, 9*(2), 143–168.

Dolron, D. (2021). *Website*. Accessed August 25, 2021, from https://desireedolron.com

Goody, J. R. (1977). Against 'ritual': Loosely structured thoughts on a loosely defined topic. In S. F. Moore & B. G. Myerhoff (Eds.), *Secular ritual* (pp. 25–35). Van Gorum.

Gordon-Lennox, J. (2017). *Emerging ritual in secular societies: A transdisciplinary conversation*. Jessica Kingsley Publishers.

Gorman, A. (2021) *The Hill We Climb: The Amanda Gorman poem that stole the inauguration show*. The Guardian. Accessed February 21, 2021, from https://www.theguardian.com/us-news/2021/jan/20/amanda-gorman-poem-biden-inauguration-transcript

Hamilakis, Y. (2011). *Archaeologies of the senses. The Oxford handbook of the archaeology of ritual and religion* (pp. 208–225). Oxford University Press.

Harari, A., & Maillard, J. (2019). Cousins des migrants. *Le Courrier* (Genève), edition 12 mars.

Holloway, M. (2015). Ritual and meaning-making in the face of contemporary death. In *Keynote lecture at symposium: Emerging rituals in a transitioning society*. University of Humanistic Studies.

Huxley, J. (1966). A discussion on ritualization of behaviour in animals and man. *Philosophical Transactions of the Royal Society of London. Series B, Biological Sciences, 251*(772), 247–248.

Jeffers, S. (1987). *Feel the fear and do it anyway*. Fawcett Columbine.

Jonte-Pace, D. (2009). Foreword: Notes on friendship. In C. Bell (Ed.), *Ritual theory, ritual practice*. Oxford University Press..

Knottnerus, J. D. (1997). The theory of structural ritualization. *Advances in Group Processes, 14*(1), 257–279.

Knottnerus, J. D. (2016). *Ritual as a missing link: Sociology, structural ritualization theory and research*. Routledge.

Kuhn, T. (1996 [1962]). *The structure of scientific revolution*. University of Chicago Press.

Lang, M., Krátký, J., & Xygalatas, D. (2020). The role of ritual behaviour in anxiety reduction: An investigation of Marathi religious practices in Mauritius. *Philosophical Transactions of the Royal Society B, 375*(1805), 20190431.

Lenoir, F. (2012). *La Guérison du monde*. Fayard.

Levine, P. A. (2010). *In an unspoken voice*. North Atlantic Books.

Levine, P. A. (2015). Foreword. In M. Picucci (Ed.), *Ritual as resource: Energy for vibrant living*. North Atlantic Books.

Levine, P. A. (2017). Front Matter. In J. Gordon-Lennox (Ed.), *Emerging ritual in secular societies: A transdisciplinary conversation*. Jessica Kingsley Publishers.

Levine, P. A. (2018). Polyvagal theory and trauma. In S. W. Porges & D. A. Dana (Eds.), *Clinical applications of the polyvagal theory: The emergence of polyvagal-informed therapies (Norton Series on Interpersonal Neurobiology)*. WW Norton & Company.

Lindahl, J. R., Fisher, N. E., Cooper, D. J., Rosen, R. K., & Britton, W. B. (2017). The varieties of contemplative experience: A mixed-methods study of meditation-related challenges in Western Buddhists. *PLoS One, 12*(5), e0176239.

Malinowski, B. (1948 [1925]). *Magic, science, and religion*. Free Press.

Mesquita, B., Boiger, M., & De Leersnyder, J. (2016). The cultural construction of emotions. *Current Opinion in Psychology, 8*, 31–36.

Moore, S. F., & Myerhoff, B. G. (1977). *Secular ritual*. Van Gorcum.

Nilsson Stutz, L. (2014). Dialogues with the dead: Imagining Mesolithic mortuary rituals. In T. Meier & P. Tillessen (Eds.), *Archaeological imaginations of religion*. Budapest.

Porges, S. W. (2011). *The polyvagal theory: Neurophysiological foundations of emotions, attachment, communication, and self-regulation (Norton Series on Interpersonal Neurobiology)*. WW Norton & Company.

Pradervand, P. (2009). *Messages of life from death row*. Booksurge Publishing.

Sassen, S. (2014). *Expulsions: Brutality and complexity in the global economy*. Harvard University Press.

Sassen, S. (2015). At the systemic edge. *Cultural Dynamics, 27*(1), 173–181.

Scaer, R. C. (2005). *The trauma spectrum: Hidden wounds and human resiliency*. W.W. Norton & Company.

Scaer, R. C. (2012). *8 keys to body–brain balance*. W.W. Norton & Company.

Scaer, R. C. (2014 [2001]). *The body bears the burden*. The Haworth Medical Press.

Schore, A. N. (2002). Dysregulation of the right brain: A fundamental mechanism of traumatic attachment and the psychopathogenesis of posttraumatic stress disorder. *Australian & New Zealand Journal of Psychiatry, 36*(1), 9–30.

Snodgrass, J. G., Most, D. E., & Upadhyay, C. (2017). Religious ritual is good medicine for indigenous Indian conservation refugees. *Current Anthropology, 58*(2), 257–284.

Sosis, R., & Handwerker, W. P. (2011). Psalms and coping with uncertainty: Religious Israeli women's responses to the 2006 Lebanon War. *American Anthropologist, 113*(1), 40–55.

Tateo, L. (2016). Fear. In V. P. Glaveanu, L. Tanggaard, & C. Wegener (Eds.), *Creativity: A new vocabulary* (pp. 43–51). Palgrave Macmillan.

Van der Kolk, B. (2014). *The body keeps the score: Mind, brain and body in the transformation of trauma*. Penguin.

Warfield, Z. J. (2020, May). The community power issue. *YES! Magazine*. Accessed August 1, 2020, from www.yesmagazine.org/magazine-article/the-community-power-issue

Jeltje Gordon-Lennox, MDiv, is a psychotherapist trained in body-based approaches and world religions. Her research and practice are influenced by her life experiences in conflict zones on several continents, in particular her work with the International Committee of the Red Cross. She has written five practical guides on secular ritualising, two in French and three in English. This collection continues the conversation on ritual and trauma begun in *Emerging Ritual in Secular Societies: A Transdisciplinary Conversation* (2017, Jessica Kingsley Publishers). Jeltje lives with her husband and their two children in Switzerland. *Website:* gordon-lennox.ch *E-mail:* jeltje@gordon-lennox.ch

Part I
Trauma and Ritual in Other Times and Places

Deeply Human

Archaeological Traces of Rituals for Coping with Death, Adversity, and Trauma

Liv Nilsson Stutz and Aaron Jonas Stutz

Archaeology reaches beyond a time accessible through oral history and historical documents to explore the trail deep in humanity's past left by the material traces of human activity. When looking at prehistory, we see that rituals have long been a human strategy for coping with change and challenges such as death, adversity and trauma. Ritual theory that focuses on practice, embodiment and sensory experience is particularly useful in the study of these traces of ritual activity. In her work on ritual, Catherine Bell makes a critical point (Bell, 1992). She sets aside the impossible question of precisely delineating what ritual *is*—and, thus, what it is not. Instead, she shifts the focus to what ritual *does*. To do this she centres on the process of *ritualisation* as an experientially and socially transformative practice. People ritualise their activities by drawing from a range of practical and communicative strategies. Through formalism and performative invariance, traditionalism, rule-governance, sacred symbolism, and self-aware performance, even acts that may be entirely habitual, entirely unreflected upon in everyday life—from food preparation and eating to social greetings, work tasks, travel, and play—can become ritualised (Bell, 1992, 1997).

Ritualisation conspicuously diverts our bodily attention away from usual everyday concerns. It takes the actor through a socially and materially scaffolded escape from normal conscious attention. The ritualisation process alters and sustains our bodily affect, often by heightening excitement or focus. It thoroughly impacts our embodied cognition, priming the body to favour intense emotional associations with particular perceptual or narrative experiences. The result may lead to affective

L. Nilsson Stutz (✉)
Department of Cultural Sciences, Linnaeus University, Växjö, Sweden
e-mail: liv.nilssonstutz@lnu.se

A. J. Stutz
Bohusläns Museum, Uddevalla, Sweden
e-mail: aaron.stutz@bohuslansmuseum.se

© The Author(s), under exclusive license to Springer Nature Switzerland AG 2022
J. Gordon-Lennox (ed.), *Coping Rituals in Fearful Times*,
https://doi.org/10.1007/978-3-030-81534-9_2

registers of calm, joy, ecstasy, or excitement and a sense of being in connection with something much bigger and more important than the self's particular or immediate concerns. What ritual does, then, is shape emotional associations with transformative awareness of being part of a greater order. Yet this order is culturally constructed, often in remarkable detail. Thus, cultural anthropologists, sociologists, and scholars of religion have observed that ritualisation supports the participants' transformative experience of connecting to traditional myths, cosmological beliefs, embodied dispositions toward fundamental symbolic structures (e.g., clean versus unclean, them versus us, female versus male, alive versus dead, etc.), and social structures.

In many instances, ritualisation brings participants into what Victor Turner referred to as anti-structure, that is, a stark contrast to the everyday social order (Turner, 1969). Anti-structural ritualised settings often openly defy and transgress cultural and social boundaries, but only during a limited period of time, marking the ritual actions and settings as unmistakably special. Here, the ritualised practices appear to highlight and value what is usually considered unacceptable, but by doing so they simultaneously reinforce prevailing, everyday social structure and values. Ritualised anti-structural practices are only accepted in a limited frame, thus containing and socially proscribing normally unacceptable actions and values. This ritual dynamic is richly described in the ethnographic literature. Examples include challenges to traditional chiefly authority (Howard & Rensel, 1994), homosexual practices (Knauft, 1985), and extramarital sexual exposure (Crapanzano, 1981). These anti-structural actions symbolically mark the ritual setting. Returning to everyday life, the participants have clarified the boundaries of social structure, but only after having ritually transgressed them. By bringing out the abnormal, chaotic, and dangerous, anti-structural ritual practices can provide a limited time to 'let off steam' while controlling subversive and dangerous forces that challenge cultural order. Ritualised practices and experiences thus deepen social bonds with other group members. Through the heightened emotional states and the embodied experience of ritualised practices that legitimise and strengthen the social structure, individuals are integrated into the larger society. Seen in this light, ritual can be viewed as central to human culture and sociality.

Ritual is thus connected to social structure and values and is especially effective in imposing order and controlling chaos. Rituals of political succession deal with chaos from within the community's varied political interests. Seasonal festivals may focus on controlling chaos from wider, unpredictable environmental factors such as bad weather, crop failures, etc. Often, ritual deals with threats to order that are intimately predictable and recurring: those life-course transformations in social status that are linked to individuals' normal biological development. Children grow up to become adults. Adults age and eventually die. Since maturation and death occur with relatively high frequency, different cultures have established ritual processes that handle these changes in a controlled manner, often through complex rites of passage that address the emotional, practical, and social strains that can arise. (Interestingly, Claude Lévi-Strauss famously argued that the story of Oedipus Rex—which was dramatised by Sophocles for performance in the highly ritualised, intricate Spring Dionysia festival—incorporates essentially anti-structural tropes, which follow one after the other in the narrative: the marked abdication of parental care; the overdevelopment of the child into a dangerously able young man, even too intelligent for his own good; the failure of parents and son to recognize one another; and the descent of plague and crop failure onto Thebes, threatening the city's very survival (Lévi-Strauss, 1955).) Other less predictable uncommon events can cause adversity and trauma for individuals and communities. In these situations humans may adapt their existing rites of affliction or rites of exchange and communication (see Bell, 1997). They may also invent new ones to place threatening or traumatising events within an ontological frame that allows participants to make sense of events, achieving a sense of controlling them. Ritual clearly becomes an effective strategy to deal with change, including the adversity and trauma that can come from interpersonal (physical or psychological) violence, environmental catastrophes, or social upheaval.

In the archaeological study of past rituals, one of the greatest challenges is in recognising traces of 'ritualised and ritualising' practices (Berggren & Nilsson Stutz, 2010). A good place to start is mortuary contexts. The ethnographic record clearly shows that, while the response to death spans incredible cross-cultural diversity, all human cultures tend to respond to death with ritual. The archaeological record reveals past diversity in the intentional handling and deposition of the remains of the dead. As death is one of the most fundamental sources of chaos that ritualised practices aim to manage, archaeological evidence of handling and depositing the dead is generally evidence of mortuary ritual. But of course, ritualised practices in the past were not limited to mortuary treatment of the dead. As we explore below, other traces of ritualised practices may be identified from archaeological sites.

In this chapter we will consider the role of ritual in a broad human context. First, we investigate the role of ritual in early human prehistory and how it emerged in our biological evolutionary lineage, known in biological classification as the genus *Homo*. The second focus of the chapter will provide examples of ritualised responses to death and crisis from the archaeological record in order to underscore how people across time and cultures have deployed ritual as a strategic way to act in the face of adversity.

The Deep History of Ritual Response to Trauma and Crisis: An Evolutionary Perspective

From a comparative evolutionary perspective, humanity (including Neanderthals) stands out as the only biological lineage in which culturally learned, systematic ritual practices emerge as a strategy to handle social crises such as death. Thus, among living species today, only we, *H. sapiens*, exhibit complex ritualised practices that act to reinforce group belonging, while at the same time, ordering the group's wider material environment and handling the cadaver when a group member dies (see Bell, above). As Mary Stiner recently observed, while members of other animal species

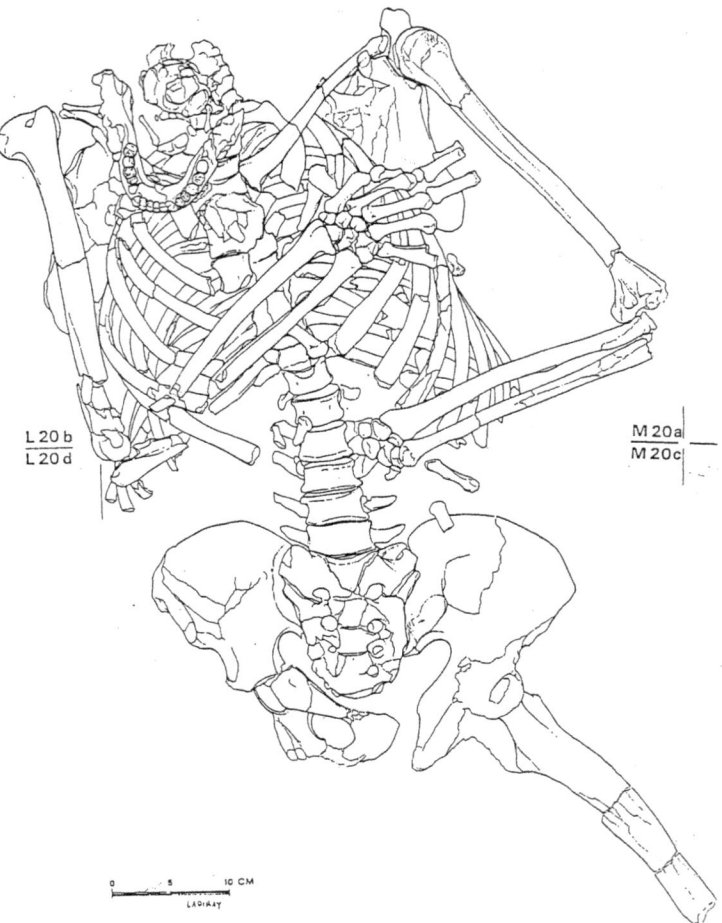

Fig. 1 Neanderthal burial, Kebara Cave, Israel. Plan view of the Neanderthal male skeleton (known as KMH 2), dated to circa 65,000–60,000 years ago; it was discovered buried in Kebara Cave in 1983. Fine-grained chemical and archaeological analyses determined that, while the legs and feet did not preserve because of natural geological conditions, the head was carefully and intentionally removed sometime after decomposition was complete. | © Kebara Archives

may *grieve* when close social relations die—and while elephants may grieve or recall the dead in a social, ritualised way—only humans invent rituals to structure the *mourning* process (Stiner, 2017) (see Fig. 1).

But when did this unique social behaviour start, and what can the deep history of ritual tell us about the role of ritual in human life and experience? It appears that ritual evolved as what evolutionary biologists call a unique-derived trait. The evolutionary ancestor we most recently share with living chimpanzees and bonobos may have engaged in ritualised greeting, aggression, hunting, territorial defence, courtship, bonding/grooming, and reconciliation behaviours. In many instances, these behaviours are learned, and they have a ritualising effect as a result of gestural and postural formalisation and invariant performance. But human ritualisation is a thoroughly evolutionarily modified social behaviour pattern, distinctive in our own natural history. It is unique in that it contributes to the emotion-laden, explicitly transformative awareness of connecting to something transcendent, which in turn is constructed socially and symbolically from the prevailing cultural context.

Indeed, many nonhuman primate species exhibit learned ritualised behaviours. Grooming, threat and aggression, submission, and reconciliation displays are likely to be part of an evolutionarily ancient adaptive system, inherited by most living primates (Stutz & Nilsson Stutz, 2018). An important physiological and biochemical aspect of primate ritualisation behaviours has to do with how individuals learn to exert self-control in the face of competing, potentially excitatory stimuli in order to maintain selective attention on a goal-oriented task. Especially in monkeys and apes, ritualised behaviours in grooming, courtship, aggression, etc. have co-evolved with self-control, which is a kind of affective auto-modulation. Yet, we suggest that, in turn, these traits have also co-evolved over nearly 35,000,000 years of monkey and ape evolution with play, which can, at least in some species and behavioural contexts, extend into adulthood.

Self-awareness in social ritualisation and play can also contribute to the emergence of theory of mind and social intelligence. A key reason that ritualised social behaviours generally contribute to collective territory defence, affiliative bonding, and conflict resolution in monkeys and apes—who constitute major primate lineages, along with lemurs, lorises, and tarsiers—is that, when compared to other mammals, primates grow up more slowly and live longer (Ross, 1998). The evolution of monkey and ape ritualised behaviour must be seen as part of a complex adaptive system that facilitated individual learning, problem solving, and memory in a complex socioecological setting, catalysing the learning of foraging and feeding behaviours, social bond formation and cooperation, social conflict solving, and reduced aggression.

Great apes—that is, orangutans, gorillas, chimpanzees, and bonobos, who count among our closest evolutionary relatives—exhibit ritualised behaviours that are important for all these socioecological purposes. Notably, different groups of chimpanzees exhibit substantial cultural variations in grooming, cooperative hunting, calling, courtship, greeting, and reconciliation behaviours (McGrew, 2004). Moreover, among wild chimpanzees in particular, individuals have occasionally been observed by primatologists to display ritualised behaviours in dealing with grief, caring for dead group members (Pettitt, 2011). These grieving behaviours emphasise selective attention and effortful interaction with the dead group member. In some instances, surviving chimpanzees interact with the cadaver for days or weeks. In one instance, a mother carried her deceased child even as decomposition and desiccation set in. She attempted to put the cadaver in familiar positions, holding on to it as if it were still living and actively able to hang on to her back. In such cases, the survivors seem aware that the cadaver is no longer a living group member, yet it is hardly an affectively neutral object for them (cf. Nilsson Stutz, 2003). Close relatives and other social relations use the still-recognizable corpse to stage familiar interactions. This may be an auto-modulative strategy for ameliorating feelings associated with anxiety, sadness, or anger. From an evolutionary perspective, it is likely that the last common ancestor to humans, chimpanzees and bonobos—living in Africa roughly seven million years ago—had evolved and constructed a very particular social and embodied niche. Ritualised behaviours were an integral part of the lived experience through imitation, social interaction, and intergenerational learning.

Ritual in Early Humans: Managing Separation from the Dead

Human evolution's point of departure was in the ecological and *phylogenetic* separation from the ape population that eventually led to chimpanzees and bonobos living in sub-Saharan Africa. In humans, cooperation has expanded from food search to food capture, transport with delayed consumption, and distribution. Human cooperation has further pervaded the teaching and learning of tool-making (Stout et al., 2019), which affords an ecological focus on learning how to extract the most nutritious and calorically rich tissues—whether from plants or from animal food resources. The complexity of human sociality has increased along with proliferation of cooperation in the emergent socially intensive niche. The potential for emotionally transformative, transcendent, and symbolically and socially structuring ritual springs from this early human evolutionary development. Thus, when looking at the evolution of unique-derived human ritualised behaviours, we can start to identify what sets human ritualisation strategies apart from those of other primates. In humans, in addition to seemingly arbitrary stereotyped or rhythmic behaviours, ritualisation involves staging and experiencing a sort of metaphorical journey, which takes the participants conspicuously away from the everyday, often to the threshold of a (culturally defined) markedly transformative, pure, or even sacred state, before returning to mundane routine (Bell, 1992). Many dimensions of human ritual's bodily and experiential impact may be the same as in the isolated chimpanzee examples of ritualised grieving. Yet, cultural construction frames these self-aware human ritual journeys within a broader ontology. Human ritualisation has thus evolved to be more complex.

We can draw key conclusions about the evolution of uniqueness in human ritual behaviours through comparative evolutionary study. Yet, the highly contentious, fundamentally archaeological question remains: When did a recognizably human pattern of ritual emerge? An early example can be found at the bottom of a 13 m deep cave shaft in Sima de los Huesos, in the Atapuerca Mountains in northern Spain. Some time between 600,000 and 300,000 years ago, the whole or partial remains of perhaps more than 28 humans (attributed to the ancient human species *Homo heidelbergensis*) were intentionally deposited or dropped into this shaft (Carbonell & Mosquera, 2006). The skeletal remains consist of mostly adolescents and young adults (10–30 years at death). Although not every individual is completely preserved, all bones of the skeleton are represented, and the bone surfaces do not exhibit any cut marks. At least some of the individuals fell into the shaft—or were dropped there—as whole cadavers, later decomposing inside the cave. It cannot be ruled out that some of the individuals may have fallen into the shaft accidentally, but the concentration of the remains suggest that the bodies of the dead were carried to the site and intentionally dropped into the shaft or left inside a now-buried entrance above the ramp (Pettitt, 2011).

The transportation of the dead to Sima de los Huesos would have been arduous work. Carrying the cadaver of an adolescent or adult human would have required the cooperation of at least two members of the community, engaging in a goal-directed activity of moving the dead to its final resting place. It is certainly possible to imagine that the bodies were accompanied by a ritualised procession, with a larger group witnessing the final event, constituted by final abandonment. As mortuary practices, these ritualised behaviours were mimetic acts (Zlatev, 2016), constructing a narrative of permanent separation from the cadaver as it dramatically disappeared into the underworld (Stutz & Nilsson Stutz, 2018). Even though researchers acknowledge great uncertainty about the language capacities of *Homo heidelbergensis*, it may be agreed that concerted, ritualised action would have contributed to the construction of complex, shared narrative experiences about mourning and separation. The cooperative ritualised handling and separation of the dead would have been a highly evocative experience, filled with episodic memories that indexed the death and loss of a relatively young group member to shared sensory experiences and embodied narratives (Nilsson Stutz, 2003). Here, ritual participation functioned as an implicit, embodied proposition about continuing one's social life, with the awareness that a kin member, ally, or antagonist had actually died.

The Emergence of Ritual for Producing Social Memory and Meaning

In the following periods, the archaeological record of mortuary practices gradually becomes richer. Burial of the dead was a relatively widespread practice among Neanderthal and early modern human populations. The oldest well-documented burials are the single primary graves from what archaeologists refer to as *Middle Palaeolithic* layers in Qafzeh and Skhul Caves, Israel. More specifically, these archaeological layers contain distinctive stone tools, traces of campfires, and hunted animal bone fragments. In particular, Qafzeh Cave—located just outside of modern Nazareth—appears to have been a favoured location to camp for Middle Palaeolithic hunter-gatherers (Hovers, 2009; Vandermeersch & Bar-Yosef, 2019). Both sites have been dated to roughly 100,000 years ago (Hovers & Belfer-Cohen, 2013; Vandermeersch & Bar-Yosef, 2019). The graves themselves consist of inhumations in simple, relatively shallow pits. But sometimes, their boundaries are marked with blocks and slabs (e.g., Qafzeh 11; see Vandermeersch & Bar-Yosef, 2019).

Only slightly later, a total of circa 40 partial skeletons provide evidence for intentional burial of the dead at many Neanderthal sites, dating to circa 80,000 to 40,000 BP, throughout Europe and the Middle East (Pettitt, 2011; Stiner, 2017). In this period, we see the manifestation of emerging, complex mortuary practices. At Kebara, Israel, the initial inhumation of a Neanderthal male was followed by careful and intentional removal of the cranium after decomposition of the soft tissues, leaving the mandible, hyoid bone, and one isolated upper molar behind in the grave (Bar-Yosef et al., 1992, 2019; Duday et al., 1990; Stutz & Nilsson Stutz, 2018). Occurring around 65,000–60,000 years ago, this ritualised treatment of an adult individual became a drawn-out engagement with the dead body, throughout its entire postmortem transformation, from cadaver to skeleton (see Fig. 1). The living curated details of the grave in memory. They used their knowledge about the postmortem decomposition of the body, planning to carefully remove the cranium, but only after a time sufficient for ligament decomposition to reach an advanced stage (Bar-Yosef et al., 2019). While skull removal as a practice is known in other sites from much more recent periods, the Kebara case is currently the oldest documented example of this mortuary treatment. Because the cranium has not been found, we can only speculate on the role of cranium removal in such early hunter-gatherer ritual practices. It is tantalising that an isolated, partially preserved cranium—albeit one with skeletal features more similar to those of anatomically modern humans from Skhul, Qafzeh, and sites throughout Africa—was intentionally placed on a remote natural shelf, deep inside Manot Cave, located 58 kilometres NNE from Kebara, sometime between circa 60,000–50,000 years ago (Hershkovitz et al., 2015).

Evidence for mortuary ritual preceding the Upper Palaeolithic or Later Stone Age periods (that is, <45,000 years ago) is mainly restricted to caves and rockshelters in western Eurasia (Hovers & Belfer-Cohen, 2013; Pettitt, 2011). This is likely a natural preservation bias imposed by geological conditions. If humans ritually separated from their dead through burial, creating commemorative places in the Near East and Europe around 100,000 to 40,000 years ago, burials were likely carried out at the same time in Africa and Eastern Asia as well. Due to the vagaries of archaeological preservation, we find many more burials, located in more varied sites, in more recent prehistoric periods. While many Upper Palaeolithic burials consist of relatively shallow pits and simple inhumations, some burials reveal spectacular ritual treatment of the dead (Riel-Salvatore & Gravel-Miguel, 2013). At the Sungir site, on the Russian Plain 200 kilometres east of Moscow, at least seven individuals were buried around 34,000–30,000 years ago, receiving a variation of treatments. Grave 2 contained two individuals, one adolescent and one juvenile, buried lying supine, head to head, their bodies covered in ochre, with thousands of mammoth ivory beads apparently sewn onto their clothes, each with ivory arm bands. One of them wore a belt with 250 fox canines. They were buried with 16 mammoth ivory spears and other items in ivory and antlers. Included in the grave was also a modified femoral shaft of an adult human. The two individuals exhibit skeletal pathologies. In at least one case, the pathology would have been clearly visible in life. In Central and Eastern Europe, in particular, individuals with marked degenerative or developmental abnormalities are relatively common in the burial record for this period (Trinkaus & Buzhilova, 2018). Much must be left to speculation. However, we can tentatively conclude that such spectacular or unusual Upper Palaeolithic burials occurred in relatively wealthy hunter-gatherer societies, where the ritualised handling of death and the dead focused not simply on separation but also—in the face of the crisis of death—on the management of power, special ritual knowledge and valuable goods.

The Evolution of Mortuary Ritual in Palaeolithic Humans: From Overcoming Separation to Strengthening Social Order and Identity

More broadly, we can follow a very gradual biocultural development in ritual as a dramatic, metaphoric journey from the everyday. At Sima de los Huesos, circa 500,000 years ago, the ritual response to death and the dead was likely intimate in social scope, perhaps only involving a few survivors, who mimetically produced and experienced a narrative of separation. Importantly, this ritual appears to have been repeated many times at the same place. Yet, the focus was directly on separation itself. It had to do with ritually and cooperatively dealing with the grief of losing a member of the group. At Sima de los Huesos, humans did not camp near the place of separation (Carbonell & Mosquera, 2006). While human language may have already become relatively complex in terms of syntax, lexical richness and non-linguistic gestures, we emphasise that joint attention and embodied experience would have been sufficient to mobilise, carry out, and make sense of the ritual. Later, with the Middle Palaeolithic burials in the Near East and Europe, from circa 100,000 to 45,000 years ago, the ritualised inhumations involved more fixed, order-dependent steps and—very likely—more cooperation. With inhumation, ritualised separation from the dead was not necessarily separation from the physical remains of the dead. In other words, interment within hunter-gatherer campsites would have meant emotional associations with places where the living continued—at least occasionally—to carry out mundane tasks, trips, and social interactions. Regardless of any intentional, reflected motivation for digging the pit, placing the body, and burying it, the practice of interment would have had a ritualising effect on the participants, further collectively linking memories of the dead or notions of death with particular places. By 100,000 years ago and onward, ritual had become a means of commemoration to produce greater social cohesion. Ritual offered a strategy for achieving social resilience when a group member died. Finally, by 30,000 years ago, in the Upper Palaeolithic period, at sites such as Sungir, ritual further became a strategy for managing the social crisis of death when individuals from wealthy families or those with special statuses or knowledge died.

The archaeological record of early mortuary ritual practices confirms that, in general, human ritual is intimately linked to human sociality and culture, offering a joint strategy for dealing with chaos and change. In the rest of this chapter we will show examples from a range of more recent archaeological contexts, allowing us to explore more extensively how human societies have employed ritual to construct cultural order in the face of crises large and small, predictable and impossibly unexpected.

Ritual as a Strategy to Handle Death

Our first archaeological examples of rituals deployed to deal with adversity in the deep past were connected to how humans have handled death. Death is a complex universal experience that can be understood as a form of crisis. On the one hand, it is mundane in the sense that it happens regularly. The average individual experiences the deaths of others many times over a lifetime. On the other hand, even an expected and 'natural' death is often deeply felt, affecting the dying individual and those closely socially or symbolically connected to her in a profoundly emotional, often lasting way (Tarlow, 2000). Death involves the loss of a social individual. It thus affects society as a whole, on a structural level. Death constantly challenges social order, as it not only reminds us of our own mortality but also creates tears in the social fabric that must be mended through the reassignment of roles, responsibilities, and even property. In this way, death creates a situation with potential for disorder and chaos. For the community to heal and repair social loss and emotional strain, death must be addressed. And ritual has long provided a powerful, enduring human strategy for the living to maintain or build resilience when someone dies.

In addition to the loss of a social being, death results in the emergence of a cadaver. Located between living and inert, between subject and object, the human cadaver is what Julia Kristeva has explored as an 'abject' (Kristeva, 1982). The cadaver embodies anti-structure: the transgression of cultural categories, threatening chaos (Turner, 1969). Ritualised practical engagement with the corpse may include acts such as washing it, dressing it, carrying it, placing it, maybe burning it, or witnessing its decomposition or interment. Adjacent ritualised acts may include speaking, singing, eating, crying, etc. Rituals have the potential to intensely affect the participant, creating or reinforcing strong emotional associations between the event itself and a larger framework of traditional myths, cosmological beliefs, and fundamental social and cultural structures. In mortuary ritual, the social and biological chaos of death is redefined, as separation from the liminal cadaver establishes the death as good, controlled, and acceptable.

Performing 'Good Death'

When analysing prehistoric mortuary practices, it is almost impossible for us to get at the *meaning* of the symbols used. What we can do is to reconstruct what people in the past were doing and what it would have looked and felt like, and from there, get at the ways in which death was produced and performed. The archaeological record of mortuary ritual reveals great variation—over long periods of time, between different cultures, and even within the same society and tradition—in the treatment of the body before disposal, the position of the body in the grave, the type and size of monuments and commemoration practices, clothing, items placed with the dead or on the grave site, etc. Within a community, however, ritualised practices for handling the dead reveal the culture's core, non-negotiable principles for producing and experiencing a good death.

The 7000-year-old burials of the Stone Age hunter-gatherer-fishers at the coastal site Skateholm in Sweden provide us with evidence of one society's enduring, non-negotiable principles for repeatedly and ritually staging the deaths of its members (Nilsson Stutz, 2003). The Skateholm site complex consists of several habitation locales and three burial grounds (one of which was completely destroyed in the 1930s and never excavated). People who lived at Skateholm buried their dead in well-defined cemeteries. A total of 85 individuals were buried there. But the inhabitants also dwelled in the surrounding southern Scandinavian landscape, sustaining themselves by hunting in the nearby dense forests while also gathering herbs, berries, seeds, and nuts. They also fished in the sea. The two preserved burial grounds were each located on small islands in a shallow lagoon. Over the course of their use, the sea level gradually rose. As one of the islands was submerged, the burial activity moved to the second island, which in turn had to be abandoned for the same reason several hundred years later.

The mortuary practices at Skateholm mainly involved inhumation, showing much variation in how the dead body was positioned and in what grave goods were placed in the grave. The dead were buried sitting or lying on the back or the side, sometimes alone and sometimes in double graves. The items placed with them ranged from stone and bone tools to beads fashioned from animal teeth, antlers, and other seemingly symbolic items.

At first glance, this variation challenges our ability to understand what was non-negotiable in the southern Scandinavian Stone Age hunter-gatherer-fisher's sense of a good death. Nonetheless, careful archaeological analysis of how the bodies of the dead were treated and placed in the graves reveals a clear pattern. Regardless of how the body was positioned in the grave, the ritualised treatment of the dead emphasised a respect for the integrity of the body. The cadavers were placed in their graves in life-like positions. When there was more than one individual in a grave, the bodies were often arranged in a way that staged them as life-like: holding each other, or looking at each other (see Fig. 2). The last image that the mourners would have had of their loved ones at Skateholm was one that staged the dead body as if it were still a living person. This last image must have resonated with deeper

Fig. 2 Young male buried with child, Skateholm I, Sweden. At this Stone Age cemetery, a young male was buried holding a child in his arms. The arrangement of the two bodies in such life-like positions suggests that their human relationship extends beyond death. | © Lars Larsson

cultural concerns and myths about life, death, the integrity of the body and the self, perhaps even individuality within a broader structure of shared humanity. It is interesting to note that while this treatment of the dead human body would have contrasted radically from the treatment of most animal cadavers, several dogs were also buried at Skateholm. Some of them received burials fully similar to those of the humans. The death and ritualised burial of a dog may sometimes have been very important to the hunter-gatherer-fisher—nearly or just as important as the death and burial of a human.

Some exceptions at the site reveal an inner tension in the Stone Age burial practices at Skateholm. While virtually all individuals were buried in order to stage a final life-like image of the dead, in one case an incomplete body was buried inside of a sack. It is possible that this treatment, which effectively concealed the state of the body, was an improvised solution to deal with an unusual or even bad death (resulting in the partition of the body). Alternatively, this extraordinary treatment may have singled this individual out in death as especially powerful, or perhaps it was a punishment for a transgression that extended beyond the grave, depriving the person of a culturally sanctioned good death. In another case, several bones from the arm and leg were carefully extracted from the grave after the process of decomposition was advanced or even terminated. This indicates that while not played up in the mortuary ritual, the processes of postmortem decay and decomposition were well understood and could be manipulated in the ritualised response to death. Beyond visually presenting a good death prior to inhumation—and occasionally hiding or depriving a dead individual of a good death—the rituals as social acts may have later incorporated the inert skeletal remains of the dead. As with head removal in Kebara Cave (see above), it is not clear what was done with the carefully and intentionally removed bones at Skateholm.

In revealing a virtually non-negotiable norm for ritually staging death, the archaeological study of Skateholm's Stone Age cemeteries allow us to see how mortuary ritual would have been used as a repeated strategy by these hunters and gatherers to deal with death and loss, as the ritualised practices would have framed death within a larger system of cultural order (Nilsson Stutz, 2003).

Deviating from the Plan: Handling Mass Death

If the burials at Skateholm are an example of how rituals would have created a good death—providing the survivors with an emotionally and socially engaging strategy to gain a sense of control over an entirely natural, normal part of the life course—archaeology can also give us insight into what happens when mass death from epidemic, natural catastrophe, or violence strains the culturally structured ritual system. In such instances, the sheer number of deaths can make the traditionally prescribed ritual response difficult or even impossible.

Historic examples of crisis deaths from the European plagues provide insights into how societies negotiated the ritual response (Bianucci & Kacki, 2012). We tend to associate plagues with mass graves. In Medieval Europe, mass graves were a dramatic break from prescribed funerary rituals involving inhumations for individuals who had died. The mass grave was a necessarily practical solution to an almost insurmountable problem. The 'Great Plague' that ravaged Provence and Languedoc during 1720–1722 claimed 119,811 deaths (approximately 30 per cent of the total population). Dozens, even hundreds of people died daily in Marseille alone. Under the circumstances, the dead were dumped in large pits and trenches that accommodated both those who died in infirmaries and were buried shortly after death and those who died elsewhere and were found hours or days later. Excavations have revealed details about the management of the dead bodies (Bianucci & Kacki, 2012). The first category was given a burial according to imposed sanitary restrictions: undressed and wrapped in shrouds. The second category was buried fully dressed to minimise contact with their bodies. The details revealed by the excavations of these pits allow us to imagine the concrete work associated with preparing and transporting the corpses. It is likely that under the circumstances the ritualised component of these burials was minimised. After all, the mass graves were a final resort.

In fact, the archaeological record of the Plague indicates that in rural areas, where the mortality was generally lower, the victims were often buried in individual graves in local cemeteries. The dead were placed on their backs, with the head oriented to the East, reflecting traditional Christian funerary treatment. It was only after transmission started to intensify that we see an adaptation to accommodate handling so many dead, first with double graves and then eventually with multiple graves (Bianucci & Kacki, 2012). Examples of this can be found at Puy-Saint-Pierre (Hautes Alpes) during the plague of 1629–1631 and in the Black Death cemeteries at Vilarnau (Passarius et al., 2008) and Saint-Laurent-de-la-Cabrerisse (Haensch et al., 2010; Kacki et al., 2011). With each epidemic we can see how the surviving population strives to maintain the traditional ritual response as the crisis grows, only reluctantly adapting the normally non-negotiable ritual response to death in extraordinary emergency conditions.

For our purposes here, it is interesting to reflect on the significant role the traditional burial practice might have played in villages and cities, as the deadly infection started to sweep through them, causing death and creating suffering. The shift in the mortuary practices shows flexibility, but the attempts to hold on to the traditional practices as long as possible also demonstrate the significant role these rituals must have played in comforting the community in this time of extreme adversity. The ability to continue providing the deceased with a prescribed burial may have allowed the survivors to hold on to the idea that this death was still acceptable. In contrast, the transition to mass graves, even organised ones, can be viewed as a material manifestation of a society trying to hold on to its most basic principles of order while sinking into chaos, as death becomes uncontrollable.

Weaponising Rituals

The recourse to ritual as a tool for healing and comfort at the time of death recurs systematically in the archaeological record. But in war and conflict it is not uncommon that ritual, or the withholding of it, can become weaponised. By denying one's dead enemies a prescribed burial ritual, or by using the bodies of the dead to make statements of power, the trauma of the violence and defeat can be prolonged among the living.

An example of this kind of postmortem violence can be found at the Bohemian stronghold Budeč, where a mass grave close to the Early Medieval hillfort recorded the gruesome story of the treatment of the bodies of the 60 soldiers who lost their lives in a massacre sometime between 930 and 990 (Štefan et al., 2016). The individuals were all young male adults. Their skeletons preserved perimortem fractures, evidence of violent deaths, mostly by sword. Between 10 and 24 of them show signs of decapitation, often as the result of multiple blows. These acts were not carried out by the professional precision of an executioner. It has been suggested that the treatment of the bodies could be understood as part of a public performance, with the bodies remaining unburied for some time, and with the skulls maintained as trophies or for display. Here, the withholding of the usual mortuary ritual, and, in addition, the added violence against the body, can be viewed as compounding trauma for survivors or those close to the victims.

Evidence of abandoned slain bodies also occurs in the archaeological record. On the battlefield in Kalkriese, which is the assumed location for the battle of the

Teutoburg Forest in 9 AD, archaeologists have found the remains of weapons, tools, personal equipment, medical equipment, horse gear, wagons, glass and silver vessels, and probably even furniture (Harnecker, 2008, 2011). Human remains from the battle were located in several contexts, with many having been buried in series of eight bone pits. The bones exhibit weathered surfaces, indicating that they had lain on the ground for some time before burial. It has been proposed that the slain Roman soldiers were left on the battlefield only to be buried several years later. The written record by Tacitus (Annales 1.62) reports that Germanicus and his legions 'on the spot, six years after the disaster, in grief and anger, began to bury the bones of the three legions, not a soldier knowing whether he was interring the relics of a relative or a stranger' (Großkopf et al., 2012, p. 101). The interpretation remains contested, since the bodies did not receive a treatment that would be typical for that of a Roman soldier (i.e., cremation), but the remains have been treated with a certain respect, in particular the skulls. If Germanicus was indeed responsible for this burial operation, it would have been improvised to fit the unique situation, with the ultimate goal of covering the dead with earth according to Roman tradition. Again, we can see the pivotal role of ritual through a drawn-out process, where at first the dead are denied their usually non-negotiable mortuary ritual, prolonging the trauma of the defeat. In this case, a Roman legion is mobilised to return, years after the event, in an emotionally costly and potentially risky endeavour, resulting in an improvised ritual to achieve as acceptable a death as possible.

Another case of abandonment of the dead can be seen at the ringfort Sandby Borg on the island Öland in the Baltic Sea. The archaeological analyses of the site indicate that the fort was raided and the male population massacred (Alfsdotter et al., 2018). Human remains have been found both inside and outside of the houses where they were left exposed. So far, no adult female bodies have been identified at the site, and it is possible that they were taken captive in the raid. However, the remains of children and adolescents were found inside the ringfort, and because the developing human skeleton does not exhibit significant differences in shape, we cannot know whether young females were among the victims (Alfsdotter & Kjellström, 2019).

Significant amounts of valuable objects such as gold brooches have been found on the site, indicating that the main objective of the event was to kill the inhabitants, not to take spoils of the raid. The presence of dead animals inside the ringfort indicates that the gates were closed after the massacre. The dead and their animals were locked inside, deprived of a prescribed mortuary ritual. In this case it seems likely that the leaving of the bodies unburied on the streets and in the houses was an integral part of the violence committed against the inhabitants (Alfsdotter, 2019). The fact that the bodies were never removed further suggests that the violent strategy might have worked, and that the fort was never used again. The presence of the unburied dead, and later on, the memory of them, may have made the place too dangerous to inhabit, cursed by bad death, not corrected and controlled with the appropriate rituals.

Rituals of Ongoing Life

A constant challenge for archaeology is to distinguish ritualised acts from mundane, non-ritualised ones. This is probably why the burials and rituals relating to death play such a prominent role in the literature. It is simply easier to demonstrate how and why these acts were ritualised. According to Catherine Bell (1992), there are no acts that are per se more 'ritual' than others. To the contrary, all acts can become ritualised in a practical context. For example, repeated and conspicuous acts, such as the deposition of horse crania or ceramic bowls in medieval house foundations in southern Scandinavia (Falk, 2008) can be viewed as rites of exchange and communication, reinforcing social roles in the household, while at the same time protecting the new house against evil, literally laying the foundation for prosperity and good fortune. Even less conspicuous yet remarkable acts, such as the deposition of natural stones, bones, and everyday tools in a fen over the course of hundreds, even thousands of years, as at Hindby Fen—studied in detail by archaeologist Åsa Berggren (2010)—may be interpreted as formally invariant, ritualised acts carried out in a marginal, special environment. Yet, their role in past social structures, connections to everyday contexts, or links to meaningful symbols are, of course, more difficult to interpret. Moreover, the relevance, motivation, emotions and symbolic associations tied to the act of making a fen offering likely changed over time, as the societies and cultural contexts of which they were part also changed (Berggren, 2010). They may have been a form of rite of communication, a set of low-key repetitive acts to maintain and nourish a connection to the spiritual realm, or a rite of affliction, carried out with the purpose of affecting the world, for good luck, fertility, a good harvest, or simply to keep the world in balance. In such archaeological cases, with accumulated deposits recording thousands of years of offerings, it is difficult to distinguish when such a formally invariant practice may have become mobilised to handle adversity or trauma, but the continued practice might also be seen as a constantly deployed strategy for avoiding events that would cause stress and hardship (Nilsson Stutz, 2016).

Ritual Responses to Catastrophic Events

Archaeology can sometimes reveal the ritual response to catastrophic natural events, such as earthquakes, tsunamis, and volcanic eruptions. In prehistoric and early historic communities, such events would have had an enormous impact on almost all aspects of life, both long and short term, as lives, possessions, resources, and familiar landscapes were lost. Essential functions of society could be destroyed. Jan Driessen has studied the impact of the Santorini volcanic eruption, dated to the second half of the second millennium BCE, that dramatically affected the Bronze Age culture of Minoan Crete. This natural disaster directly impacted Minoan society for several generations, triggering a restructuring and decentralisation of the entire economic system, with local communities assuming greater independence from palace centres. This regional political fragmentation eventually led to internal conflict and increasing competition, paving the way for the integration of Crete into the Greek Mycenaean system (Driessen, 2002).

At its height, Minoan culture was characterised by a complex hierarchical structure depending on specialized agriculture and a heightened level of organized violence between groups, making it unstable. At around 1450 BCE—around the time of the volcanic eruption—the civilisation collapsed, with settlements burning across the island and destructive aggression targeting prestigious items and symbols of authority (Driessen, 2002). These political changes were accompanied by shifts in ritual practices. The mountaintop sanctuaries in the countryside were abandoned and monumental rural shrines lost their grandeur. Instead, cave sanctuaries became more important and ritual paraphernalia started to appear in wealthy households, perhaps as a reflection of the countryside becoming less safe. Anthropomorphic representations became more important, and interestingly, volcanic pumice was used in the rituals (e.g., as part of foundation deposits in houses and caves). A ritual deposit from Knossos even indicates the practice of cannibalism, pointing to some sort of 'crisis cult' (Driessen, 2002). There are further changes in the mortuary rituals. The collective burials of the earlier Minoan period gave way to individual burials often containing status and military equipment. That these values were signalled strongly in the production of death reflected a society where the individual statuses of competing warlords had become a central concern.

The example demonstrates that as a society undergoes organisational and institutional collapse, rituals may be adapted to manage anxiety in the face of social stress. But they may also be manipulated to support new political strategies and systems emerging from a fallen civilisation's ashes. Here, it becomes clear that ritual is always articulated with other aspects of society to achieve its emotional, cognitive, and social effectiveness. It is also clear that different scales of ritual practices, from the personal and private to the public and official, often are linked.

Conclusion

Archaeology facilitates a shift in perspective on ritual, especially on how rituals provide humans with joint tools for managing crisis and trauma. Going beyond the non-explanatory platitude that ritual is a human universal, archaeology can search for, document, and study the lasting material traces of ritualised practices, revealing that ritual is a social phenomenon that emerges in a wide range of settings to manage the chaos that inevitably and repeatedly intrudes on a human community's culturally constructed—yet bodily and emotionally experienced—world. We see not only ritual's great chronological depth, reaching back hundreds of thousands of years into our evolutionary past, but we also see how ritual may be deployed to handle anxiety over misfortune, may be weaponised to produce or exert political power, and may be adapted to cope or heal from crises. In taking us on a journey that is at once metaphorical and embodied, symbolic and very phenomenologically real, ritual gives us experiences, memories, and associations that connect us to something meaningfully bigger than ourselves, and more lasting and orderly than the everyday, while also giving us a sense of strength and resilience in tackling the challenge of being human, from daily routine to lifelong practice. An archaeological perspective gives us a glimpse into the central role that ritual has played in the emergence of humanity and human cultural diversity.

References

Alfsdotter, C. (2019). Social implications of unburied corpses from intergroup conflicts: Postmortem agency following the Sandby Borg massacre. *Cambridge Archaeological Journal, 29*(3), 427–442. https://doi.org/10.1017/S0959774319000039

Alfsdotter, C., & Kjellström, A. (2019). The Sandby Borg massacre: Interpersonal violence and the demography of the dead. *European Journal of Archaeology, 22*(2), 210–231. https://doi.org/10.1017/eaa.2018.55

Alfsdotter, C., Papmehl-Dufay, L., & Victor, H. (2018). A moment frozen in time: Evidence of a late fifth-century massacre at Sandby Borg. *Antiquity, 92*(362), 421–436. https://doi.org/10.15184/aqy.2018.21

Bar-Yosef, O., Arensburg, B., Vandermeersch, B., Belfer-Cohen, A., . . . Weiner, S. (1992). The excavations in Kebara Cave, Mt. Carmel [and comments and replies]. *Current Anthropology, 33*(5), 497–550.

Bar-Yosef, O., Meignen, L., Vandermeersch, B., & Goldberg, P. (2019). The burial of the Kebara Mousterian 2nd individual. In L. Meignen & O. Bar-Yosef (Eds.), *The Middle and Upper Paleolithic archaeology of Kebara Cave, Mt. Carmel, Israel, Part II*. Peabody Museum of Archaeology and Ethnology, Harvard University.

Bell, C. M. (1992). *Ritual theory, ritual practice*. Oxford University Press.

Bell, C. M. (1997). *Ritual: Perspectives and dimensions*. Oxford University Press.

Berggren, Å. (2010). *Med kärret som källa: Om begreppen offer och ritual inom arkeologin*. Nordic Academic Press.

Berggren, Å., & Nilsson Stutz, L. (2010). From spectator to critic and participant: A new role for archaeology in ritual studies. *Journal of Social Archaeology, 10*(2), 171–197. https://doi.org/10.1177/1469605310365039

Bianucci, R., & Kacki, S. (2012). The archaeology of the second plague pandemic: An overview of French funerary contexts. In M. Harbeck, K. von Heyking, & H. Schwarzberg (Eds.), *Sickness, hunger, war, and religion: Multidisciplinary perspectives* (pp. 71–74). Rachel Carson Center for Environment and Society.

Carbonell, E., & Mosquera, M. (2006). The emergence of a symbolic behaviour: The sepulchral pit of Sima de los Huesos, Sierra de Atapuerca, Burgos, Spain. *Comptes Rendus Palevol, 5*(1–2), 155–160. https://doi.org/10.1016/j.crpv.2005.11.010

Crapanzano, V. (1981). Rite of return: Circumcision in Morocco. *The Psychoanalytic Study of Society, 9*, 15–36.

Driessen, J. (2002). Towards an archaeology of crisis: Santorini eruption and Minoan Crete. In J. Grattan & R. Torrence (Eds.), *Natural disasters and cultural change* (pp. 250–263). Routledge.

Duday, H., Courtaud, P., Crubézy, E., Sellier, P., & Tillier, A.-M. (1990). L'Anthropologie 'de terrain': Reconnaissance et interprétation des gestes funéraires. In E. Crubézy, H. Duday, P. Sellier, & A.-M. Tillier (Eds.), *Anthropologie et Archéologie: Dialogue sur les ensembles funéraires* (Vol. 2(3–4), pp. 29–50). CNRS.

Falk, A.-B. (2008). *En grundläggande handling. Byggnads offer och dagligt liv i medeltid.* Nordic Academic Press.

Großkopf, B., Rost, A., & Wilbers-Rost, S. (2012). The ancient battlefield at Kalkriese. In M. Harbeck, K. von Heyking, & H. Schwarzberg (Eds.), *Sickness, hunger, war, and religion: Multidisciplinary perspectives* (pp. 91–111). Rachel Carson Center for Environment and Society.

Haensch, S., Bianucci, R., Signoli, M., Rajerison, M., . . . Bramanti, B. (2010). Distinct clones of yersinia pestis caused the Black Death. *PLoS Pathogens, 6*(10), e1001134. https://doi.org/10.1371/journal.ppat.1001134

Harnecker, J. (2008). *Kalkriese 4. Katalog der römischen Funde vom Oberesch: Die Schnitte 1 bis 22.* Philipp v. Zabern.

Harnecker, J. (2011). *Kalkriese 5. Katalog der römischen Funde vom Oberesch: Die Schnitte 23 bis 39.* Philipp v. Zabern.

Hershkovitz, I., Marder, O., Ayalon, A., Bar-Matthews, M., . . . Barzilai, O. (2015). Levantine cranium from Manot Cave (Israel) foreshadows the first European modern humans. *Nature, 520* (7546), 216. https://doi.org/10.1038/nature14134

Hovers, E. (2009). *The lithic assemblages of Qafzeh Cave.* Oxford University Press.

Hovers, E., & Belfer-Cohen, A. (2013). Insights into early mortuary practices of Homo. In S. Tarlow & L. Nilsson Stutz (Eds.), *The Oxford handbook of the archaeology of death and burial* (pp. 631–642). Oxford University Press.

Howard, A., & Rensel, J. (1994). Rotuma: Interpreting a wedding. In M. Ember, C. Ember, & D. Levinson (Eds.), *Portraits of culture: Ethnographic originals.* Prentice Hall.

Kacki, S., Rahalison, L., Rajerison, M., Ferroglio, E., & Bianucci, R. (2011). Black Death in the rural cemetery of Saint-Laurent-de-la-Cabrerisse Aude-Languedoc, southern France, 14th century: Immunological evidence. *Journal of Archaeological Science, 38*(3), 581–587. https://doi.org/10.1016/j.jas.2010.10.012

Knauft, B. M. (1985). *Good company and violence: Sorcery and social action in a lowland New Guinea society.* University of California Press.

Kristeva, J. (1982). *The powers of horror.* (L.S. Roudiez, Trans.). Columbia University Press.

Lévi-Strauss, C. (1955). The structural study of myth. *The Journal of American Folklore, 68*(270), 428–444. https://doi.org/10.2307/536768

McGrew, W. C. (2004). *The cultured chimpanzee: Reflections on cultural primatology.* Cambridge University Press.

Nilsson Stutz, L. (2003). Embodied rituals and ritualized bodies: Tracing ritual practices in Late Mesolithic burials. *Acta Archaeologica Lundensia, 46.* Accessed August 25, 2013, from www.lunduniversity.lu.se/o.o.i.s?id=12683&postid=21368

Nilsson Stutz, L. (2016). The importance of 'getting it right': Tracing anxiety in Mesolithic burials. In J. Fleisher & N. Norman (Eds.), *The archaeology of anxiety: The materiality of anxiousness, worry, and fear* (pp. 21–40). Springer Science & Business Media.

Passarius, O., Donat, R., & Catafau, A. (2008). *Vilarnau, un village du Moyen Âge en Roussillon.* Collection Archéologie Départementale.

Pettitt, P. (2011). *The Palaeolithic origins of human burial.* Routledge.

Riel-Salvatore, J., & Gravel-Miguel, C. (2013). Upper Palaeolithic mortuary practices in Eurasia. In S. Tarlow & L. Nilsson Stutz (Eds.), *The Oxford handbook of the archaeology of death and burial.* Oxford University Press.

Ross, C. (1998). Primate life histories. *Evolutionary Anthropology: Issues, News, and Reviews, 6* (2), 54–63.

Štefan, I., Stránská, P., & Vondrová, H. (2016). The archaeology of early medieval violence: The mass grave at Budeč, Czech Republic. *Antiquity, 90*(351), 759–776. https://doi.org/10.15184/aqy.2016.29

Stiner, M. C. (2017). Love and death in the Stone Age: What constitutes first evidence of mortuary treatment of the human body? *Biological Theory, 12*(4), 248–261. https://doi.org/10.1007/s13752-017-0275-5

Stout, D., Rogers, M. J., Jaeggi, A. V., & Semaw, S. (2019). Archaeology and the origins of human cumulative culture: A case study from the earliest Oldowan at Gona, Ethiopia. *Current Anthropology, 60*(3), 309–340. https://doi.org/10.1086/703173

Stutz, A. J., & Nilsson Stutz, L. (2018). Burial and ritual. In W. Trevathan (Ed.), *The international encyclopedia of biological anthropology* (pp. 1–12). Wiley. https://doi.org/10.1002/9781118584538.ieba0081

Tarlow, S. (2000). Emotion in archaeology. *Current Anthropology, 41*(5), 713–746. https://doi.org/10.1086/317404

Trinkaus, E., & Buzhilova, A. P. (2018). Diversity and differential disposal of the dead at Sunghir. *Antiquity, 92*(361), 7–21. https://doi.org/10.15184/aqy.2017.223

Turner, V. (1969). *The ritual process: Structure and anti-structure*. Transaction Publishers.

Vandermeersch, B., & Bar-Yosef, O. (2019). The Paleolithic burials at Qafzeh Cave, Israel. *PALEO. Revue d'archéologie Préhistorique, 30*–1, 256–275.

Zlatev, J. (2016). Mimesis: The role of bodily mimesis for the evolution of human culture and language. In D. Dunér & G. Sonesson (Eds.), *Human lifeworlds: The cognitive semiotics of cultural evolution* (pp. 63–82). Peter Lang Editions.

Liv Nilsson Stutz, PhD, is an archaeologist and biological anthropologist specialising in mortuary practices. She has published widely on ritual theory and burial archaeology from a range of archaeological periods and is also active in the debates on ethics, politics, repatriation, and reburial of human remains from archaeological contexts. After receiving her PhD in archaeology from Lund University, she held the position of senior lecturer in anthropology at Emory University, and is now Professor of Archaeology at the Department of Cultural Sciences at Linnaeus University. She has conducted fieldwork in Latvia and Jordan and has co-edited the *Oxford Handbook of the Archaeology of Death and Burial* (2020 [2013], Oxford University Press) with Sarah Tarlow. *E-mail:* liv.nilssonstutz@lnu.se

Aaron Jonas Stutz, PhD, is a paleoanthropologist, specialising in research spanning the sub-fields of cultural anthropology, biological anthropology, and archaeology. He currently serves as archaeologist and osteologist at Bohusläns Museum in Uddevalla, Sweden. Aaron co-leads an interdisciplinary research team with Liv Nilsson Stutz that is investigating the Early Upper Palaeolithic deposits at Mughr el-Hamamah (Caves of the Doves) in northwestern Jordan. *Website:* https://bioculturalevolution.net *E-mail:* aaron.stutz@bohuslansmuseum.se

Ancient Rituals, Contemplative Practices, and Vagal Pathways

Stephen W. Porges

As contemplative neuroscience emerges as a discipline, research is being conducted to identify the neural pathways that contribute to compassion. Paralleling these scientific explorations, clinicians in mental health disciplines are developing interventions designed to enhance compassion of others and self (Gilbert, 2009). Limiting these investigations and applications is the lack of a consensus definition of compassion. This ambiguity limits both scientific investigations of the neural pathways determining compassion and the evaluation of compassion-based therapies.

Definitions of compassion and the tools used to assess compassion vary within the literature (see Strauss et al., 2016). Compassion has been viewed as an action, a feeling, an emotion, a motivation, and a temperament. Although common themes may be extracted from the literature, no assessment tool conforms to the standards commonly employed in scientific research (Strauss et al., 2016). Without a consensus definition, researchers investigating compassion lack a toolkit that would foster scientific inquiry, and clinicians lack a metric to assess the outcome of compassion-based therapies.

In contrast to the frequent definitions of compassion as a psychological construct, this chapter proposes that compassion is an emergent process dependent on one's neurophysiological state. Consistent with this perspective, compassion cannot be investigated as a voluntary behaviour or a psychological process independent of physiological state. Thus, compassion cannot be taught through classic rules of

A version of this paper was published as Porges, S. W. (2017). Vagal pathways: Portals to Compassion. In E. M. Seppala, E. Simon-Thomas, S. L. Brown, M. C. Worline, C. D. Cameron, & J. R. Doty (Eds.), *Oxford Handbook of Compassion Science* (pp. 189–202). Oxford University Press.

S. W. Porges (✉)
Traumatic Stress Research Consortium, Kinsey Institute, Indiana University, Bloomington, IN, USA

Department of Psychiatry, University of North Carolina, Chapel Hill, NC, USA
e-mail: sporges@indiana.edu

learning, nor can it be indexed by specific neurophysiological processes, behavioural actions, or subjective experiences independent of the bidirectional (motor and sensory) communication and neural integration between peripheral physiological state and brain function. In the proposed model of compassion, physiological state functions as an intervening variable between the person who is suffering and the responses to the person, which is manifested as the subjective experiences and behavioural actions that form operational definitions of compassion.

This chapter proposes that a physiological state mediated via vagal pathways is a necessary, but not sufficient, condition for an individual to experience compassion. The vagus is a cranial nerve, which provides the major bidirectional communication between the brain and the body. The vagus is a major component of the parasympathetic branch of the autonomic nervous system. Functionally, specific vagal motor pathways are able to inhibit the reactivity of the sympathetic branch of the autonomic nervous system, while vagal sensory pathways provide a major surveillance portal between the body and the brain. In this chapter, a model is proposed that emphasises the dependence of compassion on a vagal-mediated state that supports feelings of safety, which enable feeling one's own bodily responses at a given time, while acknowledging the bodily experiences of another person. The emphasis on shifting physiological state via vagal mechanisms to experience compassion is consistent with the historic use of rituals in contemplative training.

Since this model of compassion depends on a vagal-mediated physiological state, it may be separated from other subjective experiences that have a different physiological substrate. For example, although empathy is frequently assumed to be interchangeable with compassion, the physiological state associated with empathy may differ from the physiological state associated with compassion. Empathy is frequently operationally defined as feeling someone else's pain or negative emotion (e.g., Decety & Ickes, 2009). If we deconstruct empathy from a neurobiological perspective, empathy should be associated with the activation of the sympathetic nervous system. This would occur because the autonomic response to pain is characterised by a withdrawal of vagal influences and an activation of the sympathetic nervous system. Thus, from a neurobiological perspective, compassion is not equivalent to empathy, given that compassion engages vagal pathways.

If compassion is associated with a calm vagal state, it would promote a physiological state associated with 'safety of self' that projects calmness and acceptance towards the other. Functionally, the vagal pathways are a major component of a branch of the autonomic nervous system, historically labelled the parasympathetic nervous system. A linguistic cue for the function of the parasympathetic system is in the use of 'para' in its name. Para is derived from the ancient Greek παρά meaning 'contrary' or 'against'. Thus, the parasympathetic nervous system, as suggested by its name, provides an implicit understanding of the containment of the defensive reactivity associated with the sympathetic nervous system. Consistent with this view of the containment of defensive reactions, the critical portal to express compassion would be dependent on the capacity to recruit the vagal pathways that actively inhibit sympathetic reactivity and promote a calm physiological state that projects safety and acceptance to others.

The physiological state mediated by vagal pathways is not equivalent to compassion. Rather, it is a state that promotes or facilitates feelings of safety, positive feelings towards others (e.g., Stellar et al., 2015), connectedness, and the potential to respect both the suffering and joy of others (e.g., Kok & Fredrickson, 2010).

It is through the vagal inhibition of the neurophysiological defences (hypothalamic-pituitary-adrenal and sympathetic responses) that the vagal state functionally contains the behavioural and physiological reactivity to suffering. This containment provides opportunities to witness without judgment and to subsequently be helpful in alleviating the suffering of self or other. Brain-imaging studies attempting to distinguish between empathy and compassion are consistent with the proposed state differences associated with empathy and compassion. Klimecki et al. (2014) suggest that the excessive sharing of others' negative emotions (i.e., empathy) may be maladaptive, and that compassion training dampens empathic distress and strengthens resilience. Similarly, it has been suggested that empathy involves resonating with or mirroring another's emotion in neurophysiological, peripheral physiological, and behavioural domains (for an overview, see Decety & Ickes, 2009).

A cornerstone of compassion is respecting the individual's capacity to experience their own pain. By respecting the individual's capacity to experience pain, compassion functionally allows the individual to have their experiences 'witnessed' by another without hurting the other, by empathically sharing their pain and activating the defensive sympathetic nervous system in the other. This allows the pain to be expressed without fear of negative evaluation or the potential shame that emerges from evaluation. Compassion allows and respects the other's right to 'own' their experiences. This respect of the other in itself contributes to the healing process by empowering the other and not subjugating or diminishing the value of the person's experiences of pain or loss. Compassion functionally allows the one, who is suffering, not to be defensive or experience shame while expressing their feelings of loss and pain. If we attempt to fix the problem that produced the pain and suffering without successfully expressing compassion, the intervention will disrupt the individual's process of expression by triggering behavioural and physiological defence strategies associated with a shift in physiological state, characterised by a withdrawal of vagal influences and activation of the sympathetic nervous system. Thus, compassion relies on a 'neural' platform that enables an individual to maintain and express a physiological state of safety when confronted with the pain and suffering of others.

Vagal States Are Intertwined with the History of Contemplative Practices

Throughout the history of humanity, rituals such as chants, prayers, meditation, dance, and posture have provided the behavioural platform for contemplative practices. A careful investigation of many rituals results in the discovery that the rituals are functional exercises of vagal pathways (see Table 1). Although chants, prayers, and meditation have been incorporated into formal religions, the function of these rituals may be different from that of the narratives upon which religions were based. The narratives are attempts to fulfil the human need to create meaning out of uncertainty and to understand the unknowable mysteries of the human experience in a dynamically changing and challenging world. While this assumption may be consistent with the history of the narratives that form the corpus of formalised religions, the function of rituals may be more closely related to health and personal subjective feelings of connectedness to others, and to a deity.

The documented positive effects of meditation on mental and physical health (Bohlmeijer et al., 2010; Chiesa & Serretti, 2009; Davidson et al., 2003) have stimulated an interest in contemplative practices as health-related interventions such as mindfulness-based stress reduction (e.g., Kabat-Zinn, 2003). Science is now interfacing with insights derived from historical and often ancient contemplative practices. The accumulated knowledge suggests that meditative practices not only lead to a different perspective of reality that fosters a connectedness with others expressed through feelings of compassion, but also may have positive influences on health. These observations have led to a new discipline of contemplative neuroscience that attempts to document the shift in neural regulation that occurs during contemplative practices such as meditation.

Contemplative neuroscience has focused on documenting the mechanisms through which meditation 'heals'. Thus, contemplative neuroscience assumes directional causality in which mental processes can influence and potentially optimise bodily function. This 'top-down' model emphasises mind in the mind–body relationship and assumes that 'thought' is the driving force through which meditation functions effectively. Functionally, the research has emphasised the investigation of mind–brain relationships through imaging and electrophysiological studies of brain

Table 1 The physiology of rituals. | © Stephen W. Porges

Ritual	Vagal mechanism
Chants (vocalisations)	• Laryngeal nerves • Pharyngeal nerve • Respiration (long exhalation and deep abdominal inhalation enhance function of the vagal 'brake')
Meditation (breath)	• Respiration enhances function of the vagal 'brake'
Prayer (posture shifts)	• Carotid baroreceptors recruit vagal pathways to regulate blood pressure

circuits of expert meditators (e.g., Lutz et al., 2013). Within contemplative neuroscience, investigations of the influences of meditation on the neural regulation of visceral organs have not been emphasised.

The predominant model within contemplative neuroscience, including the study of neural pathways associated with compassion, assumes a directional causality in which mental activity drives brain function. Although this directional causality has been reliably documented (i.e., mental processes reliably influence neural activity), the model is limited because it does not incorporate two intervening variables that may mediate the effectiveness and efficiency of contemplative practices. First, the model does not acknowledge the influence of context on the nervous system. Second, the model does not acknowledge the influence of peripheral physiological state on brain function. Without detailed attention to these two variables, the functional impact of contemplative practices on mental and physical health will be unpredictable. In addition, the efficiency of contemplative practices in increasing a sense of connectedness and an ability to express compassion may be compromised.

This chapter presents a model in which contemplative practices are conceptualised as methods that require, as a prerequisite, enhanced vagal regulation of biobehavioural states. Functionally, by enhancing vagal regulation, these methods efficiently promote health and may enable expansive subjective experiences related to compassion and a universal connectedness. The model proposes that specific voluntary behaviours (e.g., breathing, vocalisations, and posture shifts), which characterise ancient rituals and form the core of contemplative practices, have the potential to trigger a physiological state that fosters health and enables subjective experiences that have been the objective of contemplative practices.

The model emphasises that two well-defined and sequential antecedent conditions are necessary for the beneficial properties of contemplative practices to be experienced. First, the environment in which contemplative practices are performed needs to have physical features that are calming and soothing. Across history and cultures, contemplative practices have been performed in quiet and safe environments. There are specific neurophysiological reasons for this consistency. To survive, humans needed to identify danger and therefore detect environments and others who were either safe or dangerous. Thus, the human nervous system needed to be sensitive to features that define physical spaces, which may either trigger or dampen defensive physiological reactivity. Second, rituals involving chants, prayers, meditation, dance, and posture shifts (e.g., kneeling, falling prostrate, etc.) provide potent stimuli to our nervous system to challenge and 'exercise' the vagal pathways that down-regulate defence and promote states of calmness and stillness.

In a safe environment, when a person no longer needs to be vigilant in anticipation of danger, the nervous system tends to shift into a qualitatively and measurably different physiological 'safe' state. This 'safe' state may function as a 'neural' catalyst for subjective feelings of social connectedness and compassion. Without the appropriate contextual cues of safety, and without the body shifting into a 'safe' physiological state, attempts at contemplative practices may be ineffective, and may even promote defensive feelings focused on self-survival that promote hypervigilance and hyper-reactivity. Consistent with this premise, via personal communications, clinicians treating veterans with post-traumatic stress disorder (PTSD) have reported situations in which mindfulness techniques have triggered defensiveness.

Polyvagal Theory: Deconstructing Ancient Rituals from a Polyvagal Perspective

Polyvagal theory (Porges, 1995, 2007, 2011) provides the neurophysiological basis to explain how rituals associated with contemplative practices contribute to bodily feelings of safety, trust, and connectedness. Polyvagal theory proposes that cues of risk and safety, which are continuously monitored by the nervous system, promote either states of safety and calmness or states of vigilance toward sources of potential threat and defence. The theory assumes that mammals are on the search for safety, which, when obtained, facilitates health and social connectedness. The theory explains how the rituals associated with contemplative practices trigger physiological states that calm neural defence systems and promote feelings of safety that may lead to expressing and feeling compassion.

The human nervous system provides two paths to trigger the neural mechanisms capable of down-regulating defensiveness to enable states of calmness that support health and connectedness. One path is passive and does not require conscious awareness (see Neuroception below) while the other requires conscious volitional behaviours that trigger specific neural mechanisms that, in turn, change one's physiological state. Spontaneous positive social behaviour expressed in facial expressions and vocal intonation is dependent on the former, and optimal outcomes of contemplative practices such as meditation and chants are dependent on the latter. Features of voice (i.e., prosody—intonation of voice) and facial expression, which characterise the interactions of positive social behaviour, provide potent cues to the nervous system that down-regulate defence circuits. In contrast to the passive pathway of calming through affiliative social engagement, contemplative training is usually conducted within the context of a 'spiritual' space (e.g., quiet space with calming music) that triggers the passive pathway to promote the physiological state associated with feeling safe. Once in a safe state, the individual can be instructed to perform voluntary behaviours such as breathing, posture shifts, and vocalisations that functionally exercise the vagal circuit and that further promote, reinforce, and strengthen states of calmness. These voluntary behaviours, which we observe as rituals, directly tap into and engage vagal circuits that efficiently manipulate one's physiological state. This enables rituals to function as neural exercises of vagal pathways.

The Role of the Vagus in Bidirectional Communication

During the phylogenetic transition from ancient reptiles to mammals, the autonomic nervous system changed. In the ancient reptiles, the autonomic nervous system regulated bodily organs via two subsystems: the sympathetic nervous system and the parasympathetic nervous system. Modern reptiles share these global features. The sympathetic nervous system provided the neural pathways for visceral changes that support defensive fight and flight behaviours. This physiological adjustment to support mobilisation for self-preservation was associated with increases in heart rate and an inhibition of digestive process, which required suppression of parasympathetic (i.e., vagal) influences to the heart and the gut.

In ancient reptiles, the parasympathetic nervous system complemented the function of the sympathetic nervous system by providing reciprocal influences on visceral organs. The reptilian parasympathetic nervous system served two primary adaptive functions: (1) when not recruited as a defence system, it supported processes of health, growth, and restoration; and (2) when recruited as a defence system, it reduced metabolic activity by dampening heart rate and respiration, enabling the 'immobilised' reptiles to appear inanimate to potential predators (i.e., a 'death-feigning' response). When not under threat, the sympathetic and parasympathetic branches of the autonomic nervous system in reptiles function reciprocally (and frequently antagonistically) to simultaneously innervate the visceral organs that support bodily functions. This synergy between the two branches of the autonomic nervous systems in support of health (not defence) is maintained in mammals, but only when mammals are safe. In this safe state, the potential of the autonomic nervous system being recruited in support of defence is greatly reduced.

Most of the neural pathways of the parasympathetic nervous system travel through the vagus nerve. The vagus is a large cranial nerve that originates in the brain stem and connects visceral organs throughout the body with the brain. In contrast to the nerves that emerge from the spinal cord, the vagus connects the brain directly to bodily organs. The vagus contains both motor fibres to influence the function of visceral organs and sensory fibres to provide the brain with continuous information about the status of these organs. The flow of information between body and brain informs specific brain circuits that regulate target organs. Bidirectional communication provides a neural basis for a mind–body science, or a brain–body medicine, by providing plausible portals of intervention to correct brain dysfunction via peripheral vagal stimulation (e.g., vagal nerve stimulation for epilepsy, depression, and PTSD) and plausible explanations for exacerbation of clinical symptoms by psychological stressors, such as stress-related episodes of irritable bowel syndrome. In addition, bidirectional communication between the brain and specific visceral organs provides an anatomical basis for historical concepts of the optimal balance among physiological systems, such as Walter Cannon's homeostasis (Cannon, 1932) and Claude Bernard's internal milieu (Bernard, 1872).

Polyvagal Theory: An Overview

Polyvagal theory is a reconceptualisation of how autonomic state and behaviour interface. The theory emphasises a hierarchical relationship among components of the autonomic nervous system that evolved to support adaptive behaviours in response to the particular environmental features of safety, danger, and life threat (Porges, 2011). The theory is named 'polyvagal' to emphasise that there are two vagal circuits: an ancient vagal circuit associated with defence (immobilisation responses) and a phylogenetically newer circuit related to feeling safe and displaying spontaneous affiliative social behaviour. The theory articulates two defence systems: (1) the commonly known 'fight-or-flight' system that is associated with activation of the sympathetic nervous system, and (2) a less-known system of immobilisation and dissociation that is associated with activation of a phylogenetically more ancient vagal pathway.

The polyvagal theory describes the neural mechanisms through which physiological states communicate the experience of safety and contribute to an individual's capacity: (a) to feel safe and spontaneously approach or engage cooperatively with others; (b) to feel threatened and recruit defensive strategies; or (c) to become socially invisible by feigning death. The theory articulates how each of three phylogenetic stages, in the development of the vertebrate autonomic nervous system, is associated with a distinct and measurable autonomic subsystem. In humans, each of these three subsystems becomes activated and is expressed physiologically under specific conditions (Porges, 2009). The three autonomic subsystems are phylogenetically ordered and behaviourally linked to three general adaptive domains of behaviour: (a) social communication (e.g., facial expression, vocalisation, listening); (b) defensive strategies associated with mobilisation (e.g., fight-or-flight behaviours); and (c) defensive immobilisation (e.g., feigning death, vasovagal syncope, behavioural shutdown, and dissociation). Based on their phylogenetic emergence during the evolution of the vertebrate autonomic nervous system, these neuroanatomically based subsystems form a response hierarchy.

The polyvagal theory emphasises the distinct roles of two distinct vagal motor pathways identified in the mammalian autonomic nervous system. The vagus is a cranial nerve that exits in the brain stem and provides bidirectional communication between brain and several visceral organs. The vagus conveys (and monitors) the primary parasympathetic influence to the viscera. Most of the neural fibres in the vagus are sensory (approximately 80%). However, most interest has been directed to the motor fibres that regulate the visceral organs, including the heart and the gut. Of these motor fibres, only approximately 15% are myelinated (i.e., approximately 3% of the total vagal fibres). Myelin, a fatty coating over the neural fibre, enables faster and more tightly regulated neural control circuits. The myelinated vagal pathway to the heart is a rapidly responding component of a neural feedback system, involving the brain and heart, which rapidly adjusts the heart rate to meet challenges.

Humans, as well as other mammals, have two functionally distinct vagal circuits. One vagal circuit is phylogenetically older and unmyelinated. It originates in a brainstem area called the dorsal motor nucleus of the vagus. The other vagal circuit is uniquely mammalian and myelinated. The myelinated vagal circuit originates in a brain stem area called the nucleus ambiguus. The phylogenetically older unmyelinated vagal motor pathways are shared with most vertebrates, and, in mammals, when not recruited as a defence system, these pathways function to support health, growth, and restoration via neural regulation of subdiaphragmatic organs (i.e., internal organs below the diaphragm). The phylogenetically 'newer' myelinated vagal motor pathways, which are observed in mammals, regulate the supradiaphragmatic organs (e.g., heart and lungs). This newer vagal circuit slows the heart rate and supports states of calmness. It is this newer vagal circuit that both mediates the physiological state necessary for compassion and is functionally exercised during rituals associated with contemplative practices.

Vagal Brake: A Mechanism to Contain Emotional Reactivity

When mammals evolved, the primary vagal regulation of the heart shifted from the unmyelinated pathways originating in the dorsal motor nucleus of the vagus to include myelinated pathways originating in the nucleus ambiguus. The myelinated vagus provided a mechanism to rapidly and efficiently regulate visceral organs to foster calm prosocial behaviours and psychological and physical health. For example, the myelinated vagus functions as an active efficient brake (see Porges et al., 1996), in which rapid inhibition and disinhibition of vagal tone to the heart can rapidly calm or mobilise an individual. Moreover, the myelinated vagus actively counteracts the sympathetic nervous system's influences on the heart and dampens hypothalamic-pituitary-adrenal (HPA) axis activity (see Porges, 2001). The vagal brake can modulate visceral state, especially the sympathetic nervous system reactions that frequently accompany empathy. Functionally, regulation of the vagal brake keeps autonomic reactivity from moving into a range that supports defensive behaviours. Thus, the vagal brake enables the individual to rapidly engage and disengage with objects and other individuals, while maintaining a physiological resource that is capable of promoting self-soothing behaviours and calm states. Ancient rituals, employing breathing, vocalisations, and posture shifts, actively recruit and exercise the vagal brake to down-regulate defensive biases and to enhance positive engagement of others with feelings of compassion.

The Face–Heart Connection: The Emergence of the Social Engagement System

When the individual feels safe, two important features are expressed. First, the bodily state is regulated in an efficient manner to promote growth and restoration (e.g., visceral homeostasis). This is accomplished through an increase in the influence of myelinated vagal motor pathways on the cardiac pacemaker (sino-atrial node) to slow heart rate and inhibit the fight-or-flight mechanisms of the sympathetic nervous system. In addition, the myelinated vagal pathways dampen the stress response system of the HPA axis (e.g., cortisol) and reduce inflammation by modulating immune reactions (e.g., cytokines). Second, through the process of evolution, the brain stem nuclei that regulate the myelinated vagus became integrated with the nuclei that regulate the striated muscles of the face and head via special visceral efferent (motor) pathways. These emergent changes in neuroanatomy provide a face–heart connection in which there are mutual interactions between the vagal influences to the heart and the neural regulation of the striated muscles of the face and head. The phylogenetically novel face–heart connection provided mammals with an ability to convey their physiological state via facial expression and prosody, enabling facial expression and voice to calm physiological state (Kolacz et al., 2018; Porges, 2011; Porges & Lewis, 2010; Stewart et al., 2013).

The face–heart connection enables mammals to detect whether a conspecific is in a calm physiological state and 'safe' to approach, or is in a highly mobilised and reactive physiological state during which engagement would be dangerous. The face–heart connection concurrently enables an individual to signal 'safety' through patterns of facial expression and vocal intonation, and potentially calm an agitated conspecific to form a social relationship. When the newer mammalian vagus is optimally functioning in social interactions (i.e., inhibiting and containing the sympathetic excitation that promotes fight-or-flight behaviours), emotions are well regulated, vocal prosody is rich, and the autonomic state supports calm, spontaneous social engagement behaviours. The face–heart system is bidirectional, with the newer myelinated vagal circuit influencing social interactions and positive social interactions influencing vagal functions to optimise health, dampen stress-related physiological states, and support growth and restoration. Social communication and the ability to co-regulate another, via reciprocal social engagement systems, leads to a sense of connectedness, which is a defining feature of the human experience.

Polyvagal theory proposes that physiological state is a fundamental part, and not a correlate, of emotion and mood. The theory emphasises a bidirectional link between brain and viscera, which would explain both how thoughts can change our physiology, and how our physiological state influences our thoughts. Thus, the initiation of contemplative practices is dependent on physiological state, although the mental process defining contemplative practices will also influence physiological state. As individuals change their facial expressions, the intonation of their voices, the pattern in which they are breathing, and the shifts their posture, they are also changing their physiology, primarily through manipulating the function of the myelinated ventral vagus pathways to the heart.

Regulating the physiological state through the myelinated vagus is an implicit underlying principle of contemplative practices. However, contemplative practices, by directly exercising the vagal regulation of state, co-opt the need for social interactions to reflexively calm the practitioner (see Neuroception below) and expand the sense of connectedness from a proximal social network to an unbounded sense of oneness. Neurophysiologically, the rituals involved in contemplative practices elicit the same neural circuits that evolved with mammals to signal safety. Through our evolutionary history, these signals were usually emitted by the mother to calm her vulnerable infant. Thus, the metaphor of the mother calming the child is neurophysiologically embedded in contemplative training and practices and is frequently used in various spiritual narratives.

As we learn more about the face–heart connection, we are informed that contemplative practices may recruit this system to obtain states of calmness. This is initially accomplished sequentially, first through the passive pathway detecting features of safety in the context in which contemplative practices are typically experienced, and then through a voluntary pathway (i.e., neural exercises) that uses efficient and reliable behavioural manipulations (e.g., breathing, vocalisations, posture shifts) that we know as rituals.

The Social Egagement System: A System that Expresses and Acknowledges Emotion

The phylogenetic origin of the behaviours associated with the social engagement system is intertwined with the phylogeny of the autonomic nervous system. As the muscles of the face and head emerged as social engagement structures, a new component of the autonomic nervous system (i.e., a myelinated ventral vagus) evolved that was regulated by the nucleus ambiguus. This convergence of neural mechanisms produced an integrated social engagement system with synergistic behavioural (i.e., somatomotor) and visceral components, as well as interactions among ingestion, state regulation, and social engagement processes. The neural pathways originating in several cranial nerves that regulate the striated muscles of the face and head (i.e., special visceral efferent pathways) and the myelinated vagal fibres formed the neural substrate of the social engagement system (see Porges, 1998, 2001, 2003a).

As illustrated in Fig. 1, the somatomotor component includes the neural structures involved in social and emotional behaviours. Special visceral efferent nerves innervate striated muscles, which regulate the structures derived during embryology from the ancient gill arches (Truex & Carpenter, 1969). The social engagement system has a control component in the cortex (i.e., upper motor neurons) that regulates brain stem nuclei (i.e., lower motor neurons) to control eyelid opening (i.e., looking),

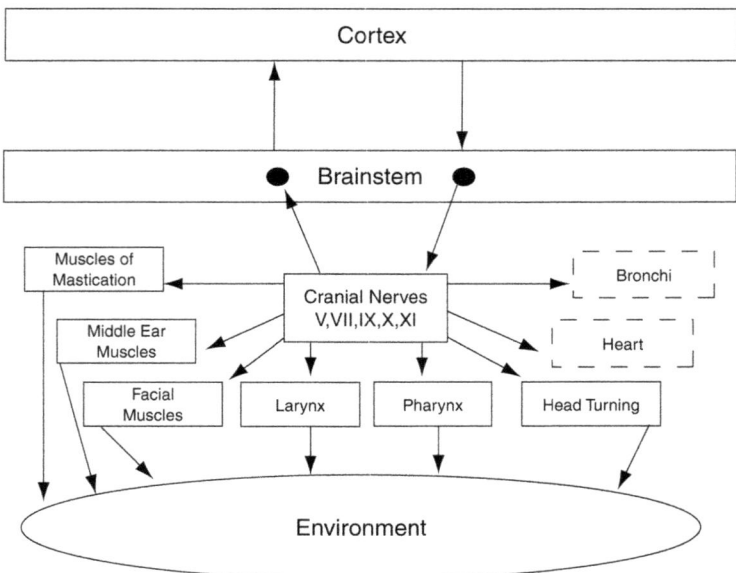

Fig. 1 The social engagement system. The social engagement system consists of a somatomotor component (i.e., special visceral efferent pathways that regulate the striated muscles of the face and head) and a visceromotor component (i.e., the myelinated vagus that regulates the heart and bronchi). Solid blocks indicate the somatomotor component. Dashed blocks indicate the visceromotor component. | © Stephen W. Porges

facial muscles (e.g., emotional expression), middle ear muscles (e.g., extracting human voice from background noise), muscles of mastication (e.g., ingestion), laryngeal and pharyngeal muscles (e.g., prosody and intonation), and head-turning muscles (e.g., social gesture and orientation). Collectively, these muscles function both as determinants of engagement with the social environment and as filters that limit social stimuli. The neural pathway involved in raising the eyelids (i.e., facial nerve) also tenses the stapedius muscle in the middle ear, which facilitates hearing human voice. Thus, the neural mechanisms for making eye contact are shared with those needed to listen to human voice. As a cluster, poor eye gaze, difficulties with extracting the human voice from background sounds, blunted facial expression, minimal head gestures, limited vocal prosody, and poor state regulation are common features of individuals with autism or psychiatric disorders.

Afferents from the target organs of the social engagement system, including the muscles of the face and head, provide potent input to the source nuclei in the brain stem regulating both the visceral and somatic components of the social engagement system. Thus, activation of the somatomotor component (e.g., listening, ingestion, lifting eyelids) could trigger visceral changes that would support social engagement, while modulation of the visceral state, depending on whether there is an increase or decrease in the influence of the myelinated vagal efferents on the sino-atrial node (i.e., increasing or decreasing the influence of the vagal brake), would either promote or impede social engagement behaviours. For example, stimulating the visceral states that promote mobilisation (i.e., fight-or-flight behaviours) will impede the ability to express social engagement behaviours.

Contemplative Practices and the Social Engagement System

The pathways defining the social engagement system enable many of the processes associated with contemplative practices (e.g., listening, chanting, breathing, shifting posture during prayer, and facial expressivity) to influence one's physiological state via a myelinated branch of the vagus. The passive pathway recruits the social engagement system (including the myelinated ventral vagus) through the cues of safety, such as a quiet environment and the presentation of prosodic vocalisations (e.g., chants) in the frequency band that would overlap with the vocal signals of safety that a mother uses to signal safety to her infant. In male-dominated religious practices, where females are not available to provide the vocal signals of safety, female-like voices are produced by boy choirs and, historically, by castrato soloists to promote feelings of spirituality.

Shifts in breathing patterns are perhaps the most accessible potent manipulations of the output of the myelinated vagus. Research documents that respiration gates the influence of the vagus on the heart (see Eckberg, 2003). The vagal inhibition of the heart's pacemaker is potentiated during exhalation and dampened during inhalation. Thus, both the duration of exhalation and the inhalation/exhalation ratio are critical in manipulating the functional 'calming' of the vagus on the heart. Rituals such as chants require extending the duration of the exhalation relative to the inhalation. Moreover, as the phrases of the chants become longer, the parameters of breathing spontaneously adjust to provide a sufficient volume of air, and breathing movements expand from the chest towards the abdomen. With abdominal or belly breathing, the diaphragm is actively pushed downward. This action stimulates vagal afferents, which functionally influence the vagal outflow to the heart. As described in Table 1, the manipulation of breathing during chants and meditation provides a potent mechanism to regulate vagal efferent activity. Thus, in these rituals, breathing strategies optimise and exercise the vagal influence on the heart.

Chants and other forms of vocalisations are frequent features of contemplative practices. These processes not only require active manipulation of breathing, but also recruit additional components of the social engagement system. For example, chants require the production and the monitoring of sounds while regulating one's breath. The modulation of vocalisations requires the active involvement of neural regulation of laryngeal and pharyngeal muscles (see Fig. 1) to change pitch and to regulate resonance. Breath is critical, since the acoustic features of vocalisations are a product of a controlled expiration, which passes air at a sufficient velocity across structures in the larynx to produce sounds.

Successful social communication via vocalisations requires rapid adjustments in both the production and detection of vocalisations. This process requires a complex feedback loop that informs brain areas of acoustic properties conveying cues of safety or danger (see Neuroception below). The cues result in dynamic adjustments in the transfer function of middle ear structures via cranial nerves to enhance or dampen the loudness of sounds within the frequency band in which social communication occurs. Without sufficient neural tone to the middle ear muscles, the sounds of human vocalisations will be lost in the low-frequency background noise that characterises our environment.

Virtually all the neural pathways involved in the social engagement system (see Fig. 1) are recruited and coordinated while chanting. This would include the regulation of muscles of the mouth, face, neck, middle ear, larynx, and pharynx. Thus, chanting may provide an efficient 'active pathway' to recruit and exercise several features of the social engagement system, while promoting a calm state through the myelinated ventral vagal pathway.

Rituals often involve voluntary posture shifts. Posture shifts influence blood pressure receptors known as baroreceptors. Baroreceptors send signals to the brainstem that will either increase heart rate by down-regulating vagal efferent output (and often stimulate sympathetic output), or decrease heart rate by increasing vagal efferent output. Manipulating posture functions as an efficient voluntary method to shift physiological state, often enabling a visceral feeling of activation (due to a transitory withdrawal of the myelinated ventral vagus) that is rapidly followed by calming (due to a reengagement of the myelinated ventral vagus).

Functionally, rituals provide a complementary alternative to social engagement behaviours, an opportunity to use voluntary behaviours to regulate and exercise several neural pathways involved in the social engagement system. As an individual becomes more proficient with the rituals, the autonomic nervous system becomes more resilient and exhibits a greater capacity to down-regulate defence and to support states that promote health, social behaviour, and compassion.

Consistent with the polyvagal theory, effective contemplative practices can only occur during states experienced as safe. Only in safe states are neurobiological defence strategies inhibited and emotional reactivity contained. Thus, a key to successful contemplative training would be to conduct contemplative exercises in an environment that supports feelings of safety. This step is mediated through the 'passive' pathway, which simultaneously down-regulates the involuntary defence subsystems and potentiates the physiological state associated with the evolutionarily newer social engagement system. Functionally, during contemplative training, the rituals involving breath, posture, and vocalisations provide, through an active pathway, 'neural' exercises of circuits involving structures described in the social engagement system. As these neural exercises enhance the efficiency and reliability of the neural pathways inhibiting defence systems, the individual acquires greater access to feelings of safety, openness, and connectedness, which are explored during contemplative practices and are antecedent states for compassion.

To understand how the 'passive' pathway is recruited, it is necessary to understand two additional features of the polyvagal theory: dissolution and neuroception.

First, through the process of dissolution (see Dissolution below), the theory describes autonomic reactivity as a phylogenetically organised response hierarchy in which evolutionarily newer circuits inhibit older circuits. Dissolution explains how specific autonomic states can support either defensive or calm behaviours. Moreover, the autonomic state that supports calm behaviour also has the capacity to actively down-regulate reactivity and defence. Thus, it is insufficient for an individual solely to abstain from defensive behaviours. The individual must also be in an autonomic state that is incompatible with defensive behaviours. Second, through the process of neuroception (see Neuroception below), context can influence one's autonomic state. Neuroception is a complex neural process that evaluates risk in the environment independently of cognitive awareness. Neuroception detects risk from sensory patterns in the environment and reflexively shifts a person's autonomic state to support either defence or safe interactions. Neuroception provides the clues to understanding how the passive pathway is elicited. Dissolution provides an understanding of the emergent hierarchical relationship among the components of the autonomic nervous system that are related to resilience and vulnerability.

Dissolution

The three circuits defined by the polyvagal theory are organised and respond to challenges in a phylogenetically determined hierarchy consistent with the Jacksonian principle of dissolution. Jackson proposed that in the brain, higher (i.e., phylogenetically newer) neural circuits inhibit lower (i.e., phylogenetically older) neural circuits and 'when the higher are suddenly rendered functionless, the lower rise in activity' (Jackson, 1882, p. 412). Although Jackson proposed dissolution to explain changes in brain function due to damage and illness, polyvagal theory proposes a similar phylogenetically ordered hierarchical model to describe the sequence of autonomic response strategies to challenges.

The human nervous system, like that of other mammals, evolved not solely to survive in safe environments, but also to promote survival in dangerous and life-threatening contexts. To accomplish this adaptive flexibility, the mammalian autonomic nervous system, in addition to the myelinated vagal pathway that is integrated into the social engagement system, retained two more primitive neural circuits to regulate defensive strategies (i.e., fight-or-flight and death-feigning behaviours). It is important to note that social behaviour, social communication, and visceral homeostasis are incompatible with the neurophysiological states that support defence. Thus, via evolution, the human nervous system retains three neural circuits, consistent with the Jacksonian principle of dissolution, that are in a phylogenetically organised hierarchy. In this hierarchy of adaptive responses, the newest circuit is used first; if that circuit fails to provide safety, the older circuits are recruited sequentially. From a contemplative practice perspective, it is necessary to recruit the phylogenetically newest circuit that down-regulates defence and involves the social engagement system and the myelinated vagus.

As described, via the active pathway, rituals exercise the integrated social engagement system, including the myelinated vagus. However, before rituals can function as efficient neural exercises, the individual must be in a calm and safe physiological state. Only in this state is the active pathway available and not in conflict with adaptive defence reactions. Thus, an understanding of how to regulate the passive pathway to maintain a calm physiological state is the initial and most critical step leading to subjective experiences related to compassion and a universal connectedness. Neuroception provides the insight into the mechanisms that enable or disable the passive pathway.

Neuroception

To effectively switch from defensive to social engagement strategies, the mammalian nervous system needs to perform two important adaptive tasks: (1) assess risk, and (2) if the environment is safe, inhibit the more primitive limbic structures involved in fight, flight, or immobilisation (e.g., death-feigning) behaviours. Any stimulus that has the potential for signalling cues of safety also has the potential to recruit an evolutionarily more advanced neural circuit that promotes calm behavioural states and supports the prosocial behaviours of the social engagement system.

The nervous system, through the processing of sensory information from the environment and from the viscera, continuously evaluates risk. Polyvagal theory proposes that the neural evaluation of risk does not require conscious awareness but functions through neural circuits that are shared with our phylogenetic ancestors. Thus, the term neuroception (Porges, 2003b, 2004) was introduced to emphasise a neural process, distinct from perception, that is capable of distinguishing environmental (and visceral) features that are safe, dangerous, or life-threatening. In safe environments, our autonomic state is adaptively regulated to dampen sympathetic activation and to protect the oxygen-dependent central nervous system, especially the cortex, from the metabolically conservative reactions of the dorsal vagal complex (e.g., fainting).

Neuroception mediates both the expression and the disruption of positive social behaviour, emotion regulation, and visceral homeostasis (Porges, 2004, 2007). Neuroception might be triggered by feature detectors involving areas of temporal cortex that communicate with the central nucleus of the amygdala and the periaqueductal gray, since limbic reactivity is modulated by temporal cortex responses to biological movements, including voices, faces, and hand gestures (Ghazanfar et al., 2005; Pelphrey et al., 2005). Embedded in the construct of neuroception is the capacity of the nervous system to react to the 'intention' of these movements and sounds. Neuroception functionally decodes and interprets the assumed goal of movements and sounds of animate and inanimate objects. This process occurs without our awareness. Although we are often unaware of the stimuli that trigger different neuroceptive responses, we are aware of our body's reactions.

Thus, the neuroception of familiar individuals and individuals with appropriately prosodic voices and warm, expressive faces translates into a positive social interaction promoting a sense of safety.

In most situations, the 'passive pathway' is activated during social interactions by identifiable social engagement features, including prosodic vocalisations, gestures, and facial expressions. However, within the proposed model, the passive pathway is recruited via exposure to the physical characteristics of the context in which contemplative training will occur. History helps us identify and describe optimal contexts. Contemplative training and practice often occur in structures with physical features that functionally remove background sounds. This contextual feature is similar to silent retreats, in which the passive triggering of 'safety' is shifted from social interactions to context. In the silent retreat, the removal of distracters, including the inhibition of potential social engagement via voice, enables the body to move from either a state of hypervigilance or a state of reciprocal interaction to a state of calmness.

Historically, structures subjectively experienced as safe were often constructed with heavy, durable materials such as stone (e.g., ancient temples). The fortress attribute supports contemplative practices through two domains: (1) protection from others when in the physically vulnerable state associated with contemplative practices; and (2) reduction of sensory cues of danger by attenuating low-frequency sounds associated with predators, and limiting distracting visual cues. In addition, the stone surfaces provided an acoustic environment in which vocalisations could be heard without effort and the acoustic characteristics were enhanced by echoes that might resonate with parts of the body. As vocalisations became ritual chants (e.g., Gregorian and Buddhist chants) the harmonics of the chants would echo through the space, and the acoustic energy would be interpreted as spiritual and healing. Physical features of these sanctuaries promote, through a passive pathway, feelings of safety and were often the contexts in which contemplative practices were taught and expressed. Thus, contemplative practices, to be functional and to have positive outcomes, must be conducted during physiological states in which the autonomic nervous system is not supporting defence and in a context that does not elicit a neuroception of danger or life threat.

Regulating Autonomic State Through Passive and Active Pathways

Within polyvagal theory, social safety depends on recruiting the ventral vagal pathways to foster a calm physiological state and maintain physiological and behavioural resilience. Consistent with compassion-focused therapy (Gilbert, 2009), the recruitment of a social safety system is a prerequisite for experiencing or expressing compassion. Neuroception describes the 'passive' pathway to recruit this state. Neuroception is the initial step to feeling safe in a safe environment.

A neuroception of safety shifts our biobehavioural state by increasing the influence of both the ventral vagus of the heart and the special visceral efferent pathways regulating the striated muscles of the face and head described in the social engagement system.

To experience a state of safety, the contextual cues in the environment have to elicit, via neuroception, the ventral vagal pathways that actively down-regulate autonomic defence systems mediated by the sympathetic nervous system and the dorsal vagus. Feeling safe requires two complementary features. First, states of hypervigilance are reduced by removing cues of distraction and potential predators. In general, the focus is on auditory and visual cues, since our nervous system is hardwired to interpret the intentionality of movements and sounds. Low-frequency sounds are hardwired cues of predators and potential life threats. High-frequency sounds are also hardwired cues of danger (see Kolacz et al., 2018; Porges & Lewis, 2010). Since our nervous system continuously attempts to interpret the intention of movements, removal of visual distracters enables individuals to shift from hypervigilance to calmness.

Removal of cues of danger is not sufficient for everyone to feel safe. Some people experience a quiet space as restful and spiritual, while others become anxious and hypervigilant. To ensure a neuroception of safety, the individual must process additional sensory features in the environment. This is most reliably accomplished through the use of acoustic stimulation that is modulated in the frequency band of a mother's lullaby. Functionally, humans are hardwired to be calmed by the modulation of the human voice (Kolacz et al., 2018; Porges & Lewis, 2010).

The acoustic features for calming infants are universal and have been repurposed by classical composers in music (Porges, 2010). Composers implicitly understood that they could lull the audience into a state of safety (i.e., via neuroception) by constructing melodic themes that duplicated the vocal range of a mother soothing her infant, while limiting the contribution of instruments that contributed low-frequency sounds. The acoustical structure of liturgical vocal music follows a similar convention by minimising low-frequency sounds and emphasising voices in the range of the nurturing mother calming her infant.

A large pipe organ, generating low-frequency tones, triggers a feeling of awe, not safety. The low tones of an organ have acoustical features that overlap with our hardwired reactions of immobilisation in the face of a predator. Thus, loud, low tones from a pipe organ could potentially disrupt the passive pathway and interfere with the state of safety required to experience compassion and a connectedness with another. However, the presentation of low-frequency tones within a confined environment may trigger a sense of submission that could be associated with psychological feelings of surrendering to a deity.

Once the passive pathway effectively shifts our physiological state, the second step can be initiated. The second step, exercising the vagal brake, recruits the 'active' (voluntary) pathway through rituals requiring manipulations of breath, posture, and vocalisations. These manipulations of the vagal brake exercise the inhibitory influence of the vagus on the heart as an efficient calming mechanism. Neurophysiologically, the vagus functions as a brake on the heart's pacemaker, resulting in the heart beating at a rate substantially slower than the intrinsic rate of the pacemaker.

Breathing is an efficient and easily accessible voluntary behaviour to systematically reduce and increase the influence of the vagus on the heart. More than a hundred years ago, Hering (1910) reported that the cardio-inhibitory vagal pathways had a respiratory rhythm that reflected the dynamic adjustment of the vagal control of the heart. Further articulation of this phenomenon was summarised as a 'respiratory gate' by Eckberg (2003) who emphasised the enhancement of the vagal influences on the heart during exhalation, and the dampening of vagal influences on the heart during inhalation. Many rituals require breathing pattern shifts. Perhaps the most obvious are chants and other forms of vocalisations, which manipulate the respiratory gate by expanding the duration of exhalation and reducing the duration of inspiration. Other rituals involving prayer and meditation may also influence vagal regulation through posture shifts, which trigger baroreceptors (blood pressure receptors) to adjust blood flow to the brain. This process involves systematic changes in vagal regulation of the heart to avoid dizziness and fainting (e.g., vasovagal syncope).

As described in Table 2, the polyvagal theory explains how the manipulation of vagal pathways is involved in the foundational processes upon which contemplative training and practice are based. These processes require two pathways (passive and active) to regulate the autonomic state and lead to a physiological state, which would enable feelings of safety and compassion to be felt and expressed. Involving the two pathways to regulate the physiological state is a prerequisite for effective contemplative practices (e.g., meditation). The two pathways function sequentially. Thus, once one is in a physiological state that supports feelings of safety, successful training would result in a resilient autonomic nervous system that would acknowledge, without mirroring, the emotional reactivity and pain often expressed by those who are suffering.

If the passive pathway does not enable the person to be in a calm ventral vagal state, then the active pathway, rather than being an enabler of compassion, may

Table 2 Steps to inner and social connectedness through effective ritualising. | © Stephen W. Porges

Step	Polyvagal process
1. Experience safe context (recruit passive pathway)	• Neuroception of safety – Remove predator cues – Add acoustic cues of a loving mother
2. Perform rituals (recruit active pathway)	• Exercise vagal brake to enhance autonomic flexibility and resilience
3. Contemplative training (e.g. specific breathing, postural, gestural and facial exercises, recitation in soothing vocal tones, meditation)	• Mental exercises involving brain functions that are dependent on maintaining 'ventral' vagal state
4. Experiencing a sense of wholeness and oneness with others or all things	• Emergent property of higher brain processes, while maintaining a 'ventral' vagal state

trigger defensiveness. If an individual engages in the active pathway in a vulnerable physiological state (during either down-regulated ventral vagal influences or up-regulated sympathetic influences), then exercising the vagal brake may create a transitory state of vulnerability. This would occur when the 'neural exercises' associated with the active pathway withdraw the vagal brake (e.g., during inspiration while meditating or chanting) and trigger a sympathetic excitation sufficient to support fight-or-flight behaviours.

Conclusion

In this chapter, a multistep sequential model is proposed that would optimise the effects of ritualising and contemplative training leading to a greater capacity to feel and express compassion. The model includes:

1. A 'passive' pathway that is elicited by feeling safe in an environment that provides sensory cues that, via neuroception, down-regulate defence;
2. An 'active' pathway that is implemented via voluntary behaviours (i.e., neural exercises of the vagal brake) capable of establishing a 'calm' neural platform (i.e., ventral vagal state) that would functionally optimise ritualising and contemplative practices;
3. Extensive contemplative training; and
4. The emergent properties of ritualising and contemplative practices, including the capacity to experience and express compassion.

The objective of this chapter is to propose that the capacity to experience and express compassion depends on a physiological state mediated by myelinated vagal pathways originating in the brain stem. Thus, within this model, the capacity to experience and express compassion is predicated on successful implementation of antecedent steps that recruit and exercise the vagal brake. Underlying this objective are several plausible assumptions and testable hypotheses:

1. Autonomic state is critical to experiencing and expressing compassion;
2. The 'passive' pathway, through neuroception, can recruit ventral vagal pathways and features of the social engagement system to shift autonomic state sufficiently to facilitate the effectiveness of rituals, as neural exercises, in enhancing autonomic regulation;
3. The 'active' pathway, through the efficient use of rituals, exercises vagal regulation of autonomic state to optimise health and resilience; and
4. The efficient use of rituals promotes a physiological state in which the outcomes of contemplative training are optimised.

Thus, an appreciation of the physiological state as an important prerequisite for compassion may result in more efficient and positive outcomes of practices, including compassion-focused therapies that would lead to enhanced compassion.

References

Bernard, C. (1872). *De la physiologie générale*. Hachette.

Bohlmeijer, E., Prenger, R., Taal, E., & Cuijpers, P. (2010). The effects of mindfulness-based stress reduction therapy on mental health of adults with a chronic medical disease: A meta-analysis. *Journal of Psychosomatic Research, 68*(6), 539–544.

Cannon, W. B. (1932). *The wisdom of the body*. W.W. Norton.

Chiesa, A., & Serretti, A. (2009). Mindfulness-based stress reduction for stress management in healthy people: A review and meta-analysis. *The Journal of Alternative and Complementary Medicine, 15*(5), 593–600.

Davidson, R. J., Kabat-Zinn, J., Schumacher, J., Rosenkranz, M., Muller, D., Santorelli, S. F., . . . Sheridan, J. F. (2003). Alterations in brain and immune function produced by mindfulness meditation. *Psychosomatic Medicine, 65*(4), 564–570.

Decety, J., & Ickes, W. J. (Eds.). (2009). *The social neuroscience of empathy*. MIT.

Eckberg, D. L. (2003). The human respiratory gate. *The Journal of Physiology, 548*(Pt 2), 339–352.

Ghazanfar, A. A., Maier, J. X., Hoffman, K. L., & Logothetis, N. K. (2005). Multisensory integration of dynamic faces and voices in rhesus monkey auditory cortex. *The Journal of Neuroscience, 25*(20), 5004–5012.

Gilbert, P. (2009). Introducing compassion-focused therapy. *Advances in Psychiatric Treatment, 15*(3), 199–208.

Hering, H. E. (1910). A functional test of heart vagi in man. *Menschen Munchen Medizinische Wochenschrift, 57*, 1931–1933.

Jackson, J. H. (1882). On some implications of dissolution of the nervous system. *Medical Press and Circular, 2*, 411–414.

Kabat-Zinn, J. (2003). Mindfulness-based interventions in context: Past, present, and future. *Clinical Psychology: Science and Practice, 10*(2), 144–156.

Klimecki, O. M., Leiberg, S., Ricard, M., & Singer, T. (2014). Differential pattern of functional brain plasticity after compassion and empathy training. *Social Cognitive and Affective Neuroscience, 9*, 873–879.

Kok, B. E., & Fredrickson, B. L. (2010). Upward spirals of the heart: Autonomic flexibility, as indexed by vagal tone, reciprocally and prospectively predicts positive emotions and social connectedness. *Biological Psychology, 85*(3), 432–436.

Kolacz, J., Lewis, G. F., & Porges, S. W. (2018). The integration of vocal communication and biobehavioral state regulation in mammals: A polyvagal hypothesis. *Handbook of Behavioral Neuroscience, 25*, 23–34.

Lutz, A., McFarlin, D. R., Perlman, D. M., Salomons, T. V., & Davidson, R. J. (2013). Altered anterior insula activation during anticipation and experience of painful stimuli in expert meditators. *NeuroImage, 64*, 538–546.

Pelphrey, K. A., Morris, J. P., Michelich, C. R., Allison, T., & McCarthy, G. (2005). Functional anatomy of biological motion perception in posterior temporal cortex: An fMRI study of eye, mouth and hand movements. *Cerebral Cortex, 15*(12), 1866–1876.

Porges, S. W. (1995). Orienting in a defensive world: Mammalian modifications of our evolutionary heritage. A polyvagal theory. *Psychophysiology, 32*(4), 301–318.

Porges, S. W. (1998). Love: An emergent property of the mammalian autonomic nervous system. *Psychoneuroendocrinology, 23*(8), 837–861.

Porges, S. W. (2001). The polyvagal theory: Phylogenetic substrates of a social nervous system. *International Journal of Psychophysiology, 42*(2), 123–146.

Porges, S. W. (2003a). The polyvagal theory: Phylogenetic contributions to social behavior. *Physiology & Behavior, 79*(3), 503–513.

Porges, S. W. (2003b). Social engagement and attachment: A phylogenetic perspective. *Roots of Mental Illness in Children, Annals of the New York Academy of Sciences, 1008*(1), 31–47.

Porges, S. W. (2004). Neuroception: A subconscious system for detecting threats and safety. *Zero to Three (J), 24*(5), 19–24.

Porges, S. W. (2007). The polyvagal perspective. *Biological Psychology, 74*(2), 116–143.

Porges, S. W. (2009). The polyvagal theory: New insights into adaptive reactions of the autonomic nervous system. *Cleveland Clinic Journal of Medicine, 76*(Suppl 2), s86–s90.

Porges, S. W. (2010). Music therapy and trauma: Insights from the polyvagal theory. In *Symposium on Music Therapy and Trauma: Bridging Theory and Clinical Practice* (pp. 3–15). Satchnote Press.

Porges, S. W. (2011). *The polyvagal theory: Neurophysiological foundations of emotions, attachment, communication, and self-regulation (Norton Series on Interpersonal Neurobiology).* W.W. Norton.

Porges, S. W., & Lewis, G. F. (2010). The polyvagal hypothesis: Common mechanisms mediating autonomic regulation, vocalizations and listening. *Handbook of Behavioural Neuroscience, 19*, 255–264.

Porges, S. W., Doussard-Roosevelt, J. A., Portales, A. L., & Greenspan, S. I. (1996). Infant regulation of the vagal 'brake' predicts child behavior problems: A psychobiological model of social behaviour. *Developmental Psychobiology, 29*(8), 697–712.

Stellar, J. E., Cohen, A., Oveis, C., & Keltner, D. (2015). Affective and physiological responses to the suffering of others: Compassion and vagal activity. *Journal of Personality and Social Psychology, 108*(4), 572–585.

Stewart, A. M., Lewis, G. F., Heilman, K. J., Davila, M. I., Coleman, D. D., Aylward, S. A., & Porges, S. W. (2013). The covariation of acoustic features of infant cries and autonomic state. *Physiology & Behavior, 120*, 203–210.

Strauss, C., Taylor, B. L., Gu, J., Kuyken, W., Baer, R., Jones, F., & Cavanagh, K. (2016). What is compassion and how can we measure it? A review of definitions and measures. *Clinical Psychology Review, 47*, 15–27.

Truex, R. C., & Carpenter, M. B. (1969). *Human neuroanatomy.* Williams and Wilkins.

Stephen W. Porges, PhD, is a distinguished university scientist at Indiana University where he directs the Traumatic Stress Research Consortium within the Kinsey Institute. He is also professor of psychiatry at the University of North Carolina. He has published more than 300 peer-reviewed papers across several disciplines including anaesthesiology, biomedical engineering, critical care medicine, ergonomics, exercise physiology, gerontology, neurology, neuroscience, obstetrics, paediatrics, psychiatry, psychology, psychometrics, space medicine, and substance abuse.

Stephen is perhaps best known for the polyvagal theory (1994), which emphasises the importance of physiological state in the expression of behavioural problems and psychiatric disorders. The theory is leading to innovative treatments, such as the music-based intervention the *Safe and Sound Protocol*™, currently used to reduce hearing sensitivities and improve spontaneous social engagement, language processing, and state regulation. *Website:* stephenporges.com *E-mail:* sporges@indiana.edu

Coping with Social Trauma in Ancient China

The Healing Power of Meditation, Ritual, and Music

Ori Tavor

Traumatic experiences can have a significant and often debilitating impact on the life of an individual, especially when they remain unresolved. The effects of a traumatic event are not limited to individuals, however; they can have long-lasting repercussions on the wellbeing of an entire community. Drawing on her work with young children who suffered trauma as a result of the conflict in Northern Ireland over the last decades of the twentieth century, Julie P. Sutton argues that a single event can have a severe impact across many levels of society. The traumatised individual will never be the same, 'nor will the immediate community, and in this way society itself will be changed' (Sutton, 2002, p. 28). Throughout most of the twentieth century, trauma was often understood to be a mental condition, and therapists have relied on talk therapy and cognitive and behavioural techniques to offer relief to their patients. Recent developments in the fields of cognitive studies and neurophysiology now offer alternative theories of trauma that define it as an activation of the body's autonomic nervous system that creates a disruption in its state of balance. These new theories have led to the creation of new therapeutic techniques designed to alleviate the corporeal effects of trauma.

In this chapter, I focus on the growing popularity of music therapy as an embodied technique designed to offset the physical, emotional, and spiritual effects of trauma. Drawing on contemporary scientific studies on the impact of music in managing emotion dysregulation associated with intrusive memories, I present key passages from early Chinese texts that discuss the transformative power of meditation, ritual, and music. Formulated during the Warring States period (481–221 BCE), one of the most tumultuous and traumatic periods in Chinese history, these works argue that participation in multimedia ritual events that combine music, dance, and scripted modes of behaviour can have a transformative effect on

O. Tavor (✉)
Department of East Asian Languages and Civilizations, University of Pennsylvania, Philadelphia, USA
e-mail: oritavor@sas.upenn.edu

the physical and emotional make-up of individuals while also bolstering group synchronisation and stimulating social conformity and compliance with social norms. These ancient Chinese texts can thus offer modern readers valuable insight into the therapeutic power of ritual and its efficacy in mitigating the effects of trauma on the individual and social body.

The Neuropsychology of Trauma

Traumatic experiences can have a debilitating impact on our daily lives. Recent years have witnessed new theories regarding the cause and impact of trauma (Scaer, 2017). Throughout the course of evolution, they claim, certain psychophysiological mechanisms developed to aid humans and animals in dealing with life-threatening events. When faced with extreme danger, the sympathetic nervous system is activated and a 'fight-or-flight' response ensues. When self-defence or a hasty retreat is not possible, the corresponding parasympathetic system, which regulates digestion and procreation, is brought into play, resulting in a state of physical paralysis. This 'freeze' state, sometimes described as 'tonic immobility', is characterised by the slowing down of pulse and blood pressure, the emptying of the gut and bladder, and a release of endorphins that numb the body to the pain of the being's imminent death.

In some cases, the activation of the parasympathetic system will result in death. In the event of survival, a 'freeze discharge' will be activated, purging the adversarial event from procedural memory and thereby restoring the natural state of balance, or 'homeostasis', between the sympathetic and parasympathetic systems. If this safety mechanism fails to take effect, the memory of the event is stored in the brain and can be triggered by similar experiences. The results of this situation are often debilitating. If the nervous system does not reset and regain balance after an overwhelming experience, it can also have adverse effects on a variety of physiological functions such as the cardiovascular, digestive, respiration, and immune systems. Unresolved physiological distress can, in turn, lead to more cognitive, emotional, and behavioural symptoms (Scaer, 2017; Levine, 2015).

Drawing on this fresh data, therapists have been developing new therapeutic techniques designed to aid patients in coping with trauma. One of these methods, developed by Peter A. Levine, is Somatic Experiencing, a technique that aims to resolve the adverse effects of trauma by guiding the patient through their internal physiological sensations rather than through their cognitive or emotional experiences. Unlike exposure therapies, which involve the direct evocation of traumatic memories, Somatic Experiencing approaches memories in a gradual and indirect manner while simultaneously fostering the creation of new corrective experiences that physically contradict those of distress and helplessness. Levine argues that the ultimate goal of this technique is 'to direct the attention of the person to internal sensations that facilitate biological completion of thwarted responses, thus leading to resolution of the trauma response and the creation of new interoceptive experiences of agency and mastery' (Payne et al., 2015, p. 15).

Much like traditional Asian techniques of meditative movement, such as Yoga, T'ai chi (Taijiquan), and Qigong, as well as various forms of seated meditation, Somatic Experiencing is designed to foster a sense of internal awareness (Payne et al., 2015). The common thread that connects these practices is their ability to stimulate specific areas of the brain.

> We know through brain-imaging techniques that specific areas of the brain 'light up' with specific activities. When a person perceives, remembers or addresses a traumatic event, the right limbic system—the part that deals with threatening experiences—'lights up', and the left prefrontal cortex (thinking brain) and Broca's area (speech expression) 'shut down'. Conversely, when we are meditating, (left frontal cortex) or verbalizing non-traumatic information (Broca's area, left frontal lobe) the right limbic system (arousal) is relatively shut down. Alternating stimulation of the left-right cerebral hemispheres, counting (left) and humming (right) hemispheres, and following a visual stimulus from right-to-left, and in-and-out are all methods of inhibiting the right limbic area. These tasks inhibit and down-regulate the amygdala through the patient/therapist bond, and the activation of both hemispheres, much like the process of attunement. (Scaer, 2017, p. 60)

Techniques such as Somatic Experiencing have grown in popularity in recent decades as alternatives to verbally intensive modes of treatment. Another regimen that has gained enthusiastic support among trauma researchers and clinicians is music therapy. Many of the current interventions used to treat cases of post-traumatic stress disorder (PTSD) among adults, such as trauma-focused cognitive behavioural therapy, Eye Movement Desensitisation and Reprocessing (EMDR), and prolonged exposure therapy, have proven effective. Unfortunately, they are also extremely time-consuming, demand a high level of clinical training, and may cause fatigue due to their intensity. Music therapy, on the other hand, is not only more accessible and less stigmatising but has also been shown to 'reduce emotional distress, foster social connectedness, and improve overall wellbeing' (Landis-Shack et al., 2017, p. 335).

Clinical studies have demonstrated that trauma is closely associated with intrusive memories that interfere with the natural state of homeostasis between the sympathetic and parasympathetic systems. Individuals suffering from post-traumatic stress often invest a great deal of time and energy into evading distressing memories, thoughts, feelings, or cues associated with their initial trauma, which can exert a dramatic negative impact on their physical, emotional, and social wellbeing. 'Through ritual, the visual, auditory, vestibular and tactile stimulation regulates, induces, promotes and establishes a state of. . .autonomic homeostasis' (Scaer, 2017, p. 56). Participation in group musical therapy is also believed to address the emotion dysregulation caused by trauma, by serving as a 'stand-in social process [designed] to address avoidant behaviour and provide positive corrective experiences' (Landis-Shack et al., 2017, p. 336). Communal music-making, after all, requires coordination and collaboration. Engaging in this project allows patients to partake in social activities in a safe space, thereby negating the feelings of isolation and distress caused by trauma. Moreover, studies have shown that music stimulates the mesolimbic dopaminergic system, an area of the brain that facilitates experiences of pleasure, reward, and arousal, and may also prompt the release of certain endorphins to the brain, boosting positive feelings while reducing negative ones. Thus, much like Levine's Somatic Experiencing, music therapy is now considered by clinicians to function as a 'resilience-enhancing intervention', a technique that 'can help trauma-exposed individuals harness their ability to recover elements of normality in their life following great adversity' (Landis-Shack et al., 2017, pp. 337–338).

Social Trauma in the Warring States

While modern therapeutic techniques such as Somatic Experiencing are based on empirical research and draw heavily on scientific terminology, the meditative exercises they utilise have been practiced in Asia for more than two millennia. The origin of the modern regimens of Taijiquan and Qigong, for example, can be traced back to a practice known as *Daoyin* (literally, 'guiding and stretching'), which combined breathing exercises, slow movement, and guided circulation of blood and *qi*[1] through the network of conduits inside the body (Kohn, 2008). In this section, I will argue that the emergence of therapeutic regimens such as *Daoyin* and seated meditation during the Warring States period in ancient China can be understood as a reaction to a state of social trauma, an attempt to restore a sense of harmony and equilibrium to both the individual and social body.

The Warring States period was one of the bloodiest and most turbulent eras in Chinese history. Following a few centuries of sociopolitical stability and economic growth under the centralised Zhou regime (est. 1045 BCE), the declining power of the royal court resulted in the fragmentation of the empire into multiple regional states led by local rulers vying for control. Motivated by the increasing brutality of everyday life, early Chinese thinkers took it as their mission to offer possible solutions to this state of chaos. Figures affiliated with the Confucian school of thought, for example, argued that the only way to regain sociopolitical stability was through a combination of moral government and an adherence to a strict regimen of individual self-cultivation. Other schools of thought, such as the Mohists and the Legalists, rejected the basic premises of Confucianism and instead advocated a system of government based on utilitarian economic principles or the mechanics of reward and punishment, respectively.

---

[1] In Traditional Chinese Medicine (TCM), *qi* is defined as the vital energy within matter that keeps it organised and makes growth possible. Within the human body, *qi* circulates in a system of channels that runs parallel to the blood vessels and is stored in major internal organs such as the heart, liver, lungs, spleen, and kidneys.

The Rise of Individual Self-Cultivation Regimens

Not all Warring States thinkers were motivated by political concerns, however. While some responded to this state of communal trauma by formulating plans to restore social order, others turned their attention to individual concerns. The Warring States philosopher Yang Zhu, for example, famously argued that he would not sacrifice even a single hair of his body for the benefit of the society. Often referred to as a radical individualist, Yang Zhu took the individual as the basis of his philosophical programme and stressed the nourishment and preservation of one's life and physical body over ethical or social concerns (Emerson, 1996).

Yang Zhu's emphasis on individual wellbeing appears in other Warring States sources as well. Recent archaeological excavations reveal the existence of a flourishing marketplace comprised of masters actively disseminating their individual self-cultivation regimens aimed at the prolongation of life and the attainment of spiritual, physical, and mental benefits (Tavor, 2016).

While some of these practices, such as *Daoyin*, involve a combination of breathing exercises with slow movements based on standardised bodily poses, others resemble what modern practitioners know as seated meditation. One of the earliest descriptions of this practice appears in the *Zhuangzi*, a late Warring States text often associated with the philosophical school of Daoism. This passage features a fictitious dialogue between Confucian and his disciple Yan Hui, who informs his master of a new self-cultivation technique he has been developing, which he calls 'sitting and forgetting'. It begins with relaxing the body and limbs and dimming sensory perception (especially sight and sound), eventually leading to a sense of transcendence from one's physical body and consciousness, and a unity between the individual and the entire universe (Mair, 1998, p. 64).

When viewed against the backdrop of recent scientific studies of trauma, 'sitting and forgetting' can be understood as a therapeutic technique aimed at purging certain adversarial events from procedural memory in order to restore the natural state of homeostasis. Living in a chaotic world filled with violence, political uncertainty, and social instability can result in a sense of helplessness from the perspective of the individual unable to control their own destiny.

Practicing seated meditation can thus facilitate the creation of a new, corrective experience that physically contradicts feelings of distress by stimulating the left hemisphere of the brain while decreasing the activity in the right hemisphere, where traumatic memories are stored, resulting in a curative state of 'forgetting'. This is further attested to in another Warring States work titled *Inward Training*. While this text is not a meditation manual, it includes multiple references to a self-cultivation technique that involves breathing exercises, sensory regulation, and attempts to control the flow of *qi* inside the body through a combination of physical and cognitive means. According to the metaphysical framework of the text, everything in the world is animated by the same basic energy: *qi*. Flowing around the universe, it lacks a fixed position, entering and exiting all living things in a manner that looks random to the occasional observer. The goal of the text is simple: to allow the practitioner to understand the patterns of *qi* circulation, giving them the power to accumulate it inside their bodies, resulting in a complete cognitive, physiological, and spiritual transformation.

The author of *Inward Training* argues that the key for the success of this endeavour is reaching a state of stillness (*jing*) and emptiness (*xu*). Sensory perception, as well as strong emotions such as excessive anger and sorrow, occupy and overstimulate our minds, preventing us from achieving a state of balance. The only solution is a technique called 'sweeping the seat of consciousness'. It begins with assuming a fixed seating position in which the limbs are squared and properly aligned, followed by a concentrated effort to reduce sensory input, eliminate desire, and cut off all mental cogitation. Once this state of stillness and emptiness is achieved, vital *qi* will enter the body of the practitioner and take up residence within it, resulting in an emotional and mental experience of oneness and harmony with the universe. This mental state is accompanied by distinct physical boons such as improved perception and physical wellbeing (Roth, 1999).

The Therapeutic Power of Ritual and Music

The meditative regimens described in the *Zhuangzi* and *Inward Training* represent an individualised therapeutic solution to the social trauma that followed the collapse of the Zhou order. Recently excavated manuscripts suggest that a significant number of educated elites responded to this situation by tending to their personal wellbeing, sometimes at the expense of public engagement. This posed a threat to Confucian thinkers, who believed that the only viable route to social and political harmony depended on the active participation of educated elites in the work of government. These sentiments are perhaps best manifested in the work of third-century Confucian thinker Xunzi, who, like his intellectual forefather Confucius (551–479 BCE) before him, was not only an educator and an advisor, but also a ritual master, a curator, and preserver of ancient rites. As a ritualist, Xunzi took it upon himself to rearticulate the Confucian project of moral self-cultivation as a therapeutic technique superior to the meditational and calisthenics regimens advocated by other Warring States masters (Tavor, 2013).

It is important to note that Xunzi did not question the efficacy of individualised therapeutic regimens in coping with trauma. His writings suggest that he was well acquainted with these practices and the philosophical and medical terminologies employed by their proponents. Xunzi's main critique was based on economic and sociopolitical grounds, in that individual self-cultivation requires a substantial financial investment (hiring the services of professional masters and purchasing their manuals) and often distracts educated elites who might otherwise employ their skills to help the entire community recuperate from the traumatic events of the Warring States period. Instead, Xunzi opted to promote his own therapeutic regimen, which allows people to gain individual bounties such as good health, sensory and emotional satisfaction, and moral edification, while at the same time promoting a sense of communal identity, enforcing social hierarchies, and maintaining political order. The core of this regimen involves participation in choreographed ritual events (*li*) that were accompanied by a musical performance (*yue*).

The last two decades have witnessed a rapid rise in cognitive and neuroscientific research on music. Biomusicologists such as Steven Brown study the evolutionary origins of music and argue that it is a cooperative device designed to enhance group survival, as it enhances the ability of individual humans to coordinate with each other and act in synchronicity. Brown asserts that recent studies show that the human brain has specific neural areas that control and regulate harmony and meter, which are also connected to two of the key cognitive systems of attention and reward. From an evolutionary perspective, we may surmise that these mechanisms developed to facilitate cooperation and coordination. Music and ritual thus evolved as cultural devices, or 'generalized emotive manipulators', which enhance the memorability of certain events, augmenting them in our collective consciousness (Brown, 2003, p. 16–17).

Music's homogenising effect and its ability to stimulate social conformity and compliance with social norms are based on two important elements. First, on a basic level, participating in musical group events such as religious ceremonies, festivals, and raves is often presented as an important criterion for membership in a community. Second, musical devices such as rhythm, repetition, and polyphony 'act to increase the meaning and memorability of linguistic messages' (Brown, 2006, p. 4). It is important to note that Brown's adaptationist views, which depict music an evolutionary mechanism that was useful to the survival of the human species, are not shared by all. Cognitive psychologists such as Aniruddh D. Patel argue that music is a biologically powerful human invention that builds on diverse, pre-existing brain functions, rather than a trait that originated via processes of natural selection and biological adaptation. Music does not change our brains on a genetic level; instead, it is a technology that 'has to be learned anew by each new generation of human minds' (Patel, 2010, p. 43).

Regardless of the exact cognitive and neurological mechanisms that enable it, the realisation that ritual and music have the power to foster group cohesion and enhance compliance to social norms and values has been widely acknowledged in ancient China and actively championed by Confucian thinkers such as Xunzi. In the 'Discourse of Ritual' chapter of the book bearing his name, Xunzi hails ritual as the perfect tool for guaranteeing sociopolitical stability, claiming that 'those under Heaven who follow it will have good order; those who do not follow it will have chaos; those who follow it will have safety; those who do not follow it will be endangered; those who follow it will be preserved; those who do not follow it will perish' (Hutton, 2014, p. 205).

In Xunzi's philosophical vision, human beings are born with innate tendencies towards selfish behaviour and an unquenchable thirst to satisfy their most basic desires for food, sex, wealth, and fame. Given the chaotic world into which Xunzi was born, two centuries into the collective social trauma of Warring States era, this bleak view of humanity is understandable. It is important to note, however, that Xunzi's acceptance of the inherent faults of human nature is not pessimistic. In fact, his philosophy is predicated on a deep belief in the human ability to change, learn, and overcome trauma in order to re-establish the harmonious society of an imagined ancient past. This can be achieved by an adherence to a life-long regimen of moral self-cultivation that draws on external social institutions and cultural devices, such as education, textual study, and participation in communal ritual events that feature choreographed musical performances.

The musical component of these philosophies is particularly important, as Warring States thinkers often associate music with the regulation of emotions. Xunzi open his 'Discourse on Music' chapter with the claim that music is an intrinsic part of human existence. 'Music is joy, an unavoidable human disposition; people cannot be without music, if they feel joy, they must express it in sound and give it shape in movement' (Hutton, 2014, p. 218). Moreover, much like Brown, Xunzi also sees music as an 'emotive manipulator' that can be used to unify a group of people in a non-coercive manner by marking certain events and augmenting them in memory. When music is performed in public, argues Xunzi, each member of the community, from the ruler to the lowliest subordinate, listens to it together, resulting an increased sense of cohesion and harmony:

> [M]usic observes a single standard in order to fix its harmony, it brings together different instruments in order to ornament its rhythm, and it combines their playing in order to achieve a beautiful pattern; it can thus lead people in a single, unified way, and is sufficient to bring order to the myriad changes within them. (Hutton, 2014, p. 218)

It is clear from these passages that Xunzi sees the participation in ritualised musical events as a key component in resolving the sense of helplessness and anxiety brought on by an extended period of war and the resulting collapse of the sociopolitical order. What makes ritual and music unique, and in some ways superior, to more verbal forms of education and governance is its embodied nature. 'The sounds of music enter into people deeply and transform them quickly,' Xunzi argues. 'It has the power to make good the hearts of people, to influence deeply, and to reform their manners and customs with facility' (Hutton, 2014, p. 219–220). Drawing on the terminology used by the proponents of the *Daoyin* calisthenics regimens and the techniques of seated meditation, Xunzi traces the efficacy of ritual and music to their ability to control the flow of blood and *qi* within the human body. Earlier in his work, he criticises those who attempt to prolong their lives using individual self-cultivation regimens and instead advocates an adherence to a strict regimen of ritual prescriptions as the superior technique for achieving a state of physical, mental, and spiritual harmony and equanimity (Hutton, 2014, p. 10).

In the chapter 'Discourse on Music' Xunxi elaborates on this statement, arguing that while lavish, excessive, and lurid musical performances cause 'perverse *qi*' to take form within one's body, participation in proper musical events produces 'compliant *qi*' that brings forth harmony within the entire social body but also induces individual physiological effects such as a balanced flow of blood and *qi* and improved sensory perception (Hutton, 2014, p. 221).

In her recent book on music in ancient China, Erica Brindley demonstrates that musical metaphors played a central role in Warring States political, philosophical, and medical discourse. Her reading of the *Xunzi* supports many of the points made earlier in this chapter, particularly her claim that Xunzi acknowledges music as a 'powerfully influential, manipulative, and fool-proof device' for eliciting certain responses in the human body (Brindley, 2012, p. 108). In addition, Brindley also draws our attention to the centrality of terms such as harmony and equilibrium in Xunzi's discussion of the physiological effects of participating in ritualised musical events. Reading his discussion on 'compliant *qi*', Brindley suggests that in Xunzi's eyes, proper music 'takes on the role in the body of what we would now describe as serotonin or endorphins' (Brindley, 2012, p. 138).

Recent clinical studies among patients who suffered a stroke seem to support the general sentiment of this claim. These studies connect music's efficacy to its ability to stimulate a variety of non-musical brain functions, enhance neuroplasticity, and most importantly, impact the limbic system. The findings of these trials suggest that regular exposure to music helps to lower the levels of the hormone cortisol, thereby improving the function of the hippocampus, a region in the brain that forms part of the limbic system and is primarily associated with memory and spatial navigation (Patel, 2010).

For Xunzi, ritualised musical performances can be seen as an aesthetic training ground, a unique mode of practice different from everyday activity. Within ritual time and space, routine gestures and movements take on special meanings precisely because they are performed outside of a regular context. Far from being a frivolous satisfaction of desire, ritual and music provide meaningful experience that induces a bodily transformation and is accompanied by a sense of pleasure greater than simple carnal joy (Tavor, 2013). Moreover, when read against the backdrop of contemporary scientific literature on trauma, we can also see that for Xunzi, the therapeutic value of ritual and music lies in their corporality and their ability to promote a message of harmony and wellbeing in a non-verbal way.

In a society plagued by violence and precariousness, participation in such events offers a respite for the harsh everyday reality and can function as a remedy to the state of social trauma by fostering cooperation and enhancing in-group harmony. At the end of the 'Discourse on Music' chapter, Xunzi offers a description of dance (*wu*), which refers to the choreographed component of the ritual event accompanied by music. He argues that this performance aims to represent the ideal pattern of a well-ordered universe. The beat of the drums represents Heaven,[2] the bells represents Earth, the stone chimes represent water, the various wind instruments represent the sun, moon, and stars, and the smaller percussion instruments represent the myriad creatures. When the dancers move their bodies to the music, they do not necessarily comprehend the idea behind it. As Xunzi puts it:

> [H]ow does one know the meaning of the dance? The eyes do not themselves see it, and the ears do not themselves hear it; nevertheless, it controls their postures, gestures, directions,

---

[2] While the term Heaven (*tian*) often refers to a high deity, in Xunzi's philosophical system it does not have religious connotations and instead indicates something akin to the phenomena and processes of the natural world.

and speed; when all the dancers are restrained and orderly exerting to their utmost the strength of their bones and sinews to match the rhythm of the drum and bell sounding together, and no one is out of step, then how easy it is to tell the meaning of this group gathering! (Hutton, 2014, p. 222)

Musical performances and ritual dancing, according to Xunzi, require a great deal of attention to detail, restrained control, and rigorous practice, and this is not something that can be learned in theory or through observation. The artists have to go through the experience by themselves, dedicate a long time to practice and rehearsal, and only then will they be able to understand the importance of it. Ancient Chinese ritual performance, much like modern applications of music therapy, are thus a means for coordinating the rhythm of one's body not only with other dancers but with the universe itself, resulting in a sense of harmony, balance, and overall wellbeing that can counteract the memories of past traumatic events. These events reshape the bodies and minds of the participants by activating non-musical brain functions and stimulating the limbic system to promote the restoration of homeostasis while enhancing group synchronicity and cohesion to counteract the effects of social trauma.

Conclusion

In her study of meditation and ritual practices, Barbara Lex demonstrates that both individual meditation and communal ritual have similar neurophysiological effects. In meditation, the reduced sensory output monopolises the left hemisphere, creating a trophotropic response—the relaxing of the muscles. In ritual, on the other hand, the repetitive physical stimuli and musical rhythms bombard the nervous system, over-take the right hemisphere, and create an ergotropic response—an increase in muscle tonus. The cumulative dominance of one hemisphere, however, causes the other to increase its own activity in compensation. Thus, while stimulating opposite sides, both communal ritual and the individual practice of meditation utilise alternating stimulation of the bihemispheric brain that prompts a neurological and physiological response (Lex, 1979).

The growing popularity of individual techniques such as the seated mediation of the *Inward Training* and the ritualised musical performances described in the *Xunzi* demonstrate that the therapeutic potential of such regimens was well known to early Chinese philosophers, who promoted them as a cure to the social trauma of the Warring States period. Threatened by the success of individual therapeutic techniques, Confucian thinkers created a new theoretical framework that emphasised the physical, cognitive, and spiritual effects of ritual and music, stressing that participation in such events can bring forth the same therapeutic benefits of meditation, namely the restoration of homeostasis, but on a much larger scale—promoting harmony and ensuring the wellbeing of the entire social body.

The therapeutic regimens that emerged during the Warring States period continued to develop and flourish throughout Chinese history, especially during periods of social trauma. It is thus hardly surprising that a variety of individual and communal practices began to resurface in the aftermath of the Cultural Revolution (1966–1976), a period that is widely believed to have left a permanent scar on the collective psyche of the Chinese nation. For example, the 1980s were marked by a resurging interest in Qigong, an individual regimen that combines seated meditation with ritualised movement, which was touted as a 'somatic science', or a secular therapeutic practice for the modern age (Palmer, 2007).

Recent years have also witnessed the revival of Confucianism and its communal ritual practices. With the support of the state, new rituals designed to mark important occasions such as births, weddings, and funerals have begun to grow in popularity (Billioud & Thoraval, 2015). Dissatisfied with the public school system, parents in China are now sending their children to traditional Confucian academies where they are exposed to a strict regimen of musical education designed to aid in their 'natural emotional, moral, and spiritual growth' while encouraging socialisation and group harmony in the face of what is believed to be a rampant individualism plaguing contemporary Chinese society (Ji, 2008, p. 113). These examples demonstrate that, far from being seen as relics of the past, the therapeutic techniques developed during the Warring States period to deal with social trauma are still considered valuable. These regimens might therefore serve as a potential source of inspiration for contemporary clinicians and ritualists searching for effective ways of coping with social trauma in the modern world.

References

Billioud, S., & Thoraval, J. (2015). *The sage and the people: The Confucian revival in China.* Oxford University Press.

Brindley, E. (2012). *Music, cosmology, and the politics of harmony in early China.* State University of New York Press.

Brown, S. (2003). Biomusicology, and three biological paradoxes about music. *Bulletin of Psychology and the Arts, 4,* 15–17.

Brown, S. (2006). How does music work? Toward a pragmatics of musical communication. In S. Brown & U. Volgsten (Eds.), *Music and manipulation: On the social uses and social control of music* (pp. 1–27). Berghahn Books.

Emerson, J. (1996). Yang Chu's discovery of the body. *Philosophy East and West, 46*(4), 533–566.

Hutton, E. (2014). *Xunzi: The complete text.* Princeton University Press.

Ji, Z. (2008). Educating through music: From an 'initiation into classical music' for children to Confucian 'self-cultivation' for university students. *China Perspectives, 75,* 107–117.

Kohn, L. (2008). *Chinese healing exercises: The tradition of Daoyin.* University of Hawai'i Press.

Landis-Shack, N., Heinz, A. J., & Bonn-Miller, M. O. (2017). Music therapy for posttraumatic stress in adults: A theoretical review. *Psychomusicology: Music, Mind, and Brain, 27*(4), 334–342.

Levine, P. A. (2015). Somatic experiencing. In E. S. Neukrug (Ed.), *The Sage encyclopedia of theory in counseling and psychotherapy* (pp. 951–953). Sage.

Lex, B. (1979). The neurobiology of ritual trance. In E. D'Aquili, C. Laughlin, & J. McManus (Eds.), *The spectrum of ritual: A biogenetic structural analysis* (pp. 117–151). Columbia University Press.

Mair, V. (1998). *Wandering on the way: Early Taoist tales and parables of Chuang Tzu*. University of Hawai'i Press.

Palmer, D. (2007). *Qigong fever: Body, science and utopia in China*. Columbia University Press.

Patel, A. D. (2010). Music, biological evolution, and the brain. In M. Bailar (Ed.), *Emerging disciplines* (pp. 91–144). Rice University Press.

Payne, P., Levine, P. A., & Crane-Godreau, M. A. (2015). Somatic experiencing: Using interoception and proprioception as core elements of trauma therapy. *Frontiers in Psychology, 6*, 1–18.

Roth, H. (1999). *Original Tao: Inward training (Nei-yeh) and the foundations of Taoist mysticism*. Columbia University Press.

Scaer, R. C. (2017). The neurophysiology of ritual and trauma: Cultural implications. In J. Gordon-Lennox (Ed.), *Emerging ritual in secular societies: A transdisciplinary conversation* (pp. 55–67). Jessica Kingsley Publishers.

Sutton, J. (2002). *Music, music therapy and trauma: International perspectives*. Jessica Kingsley Publishers.

Tavor, O. (2013). Xunzi's theory of ritual revisited: Reading ritual as corporal technology. *Dao: A Journal of Comparative Philosophy, 12*(3), 313–330.

Tavor, O. (2016). Authoring virile bodies: Self-cultivation and textual production in early China. *Studies in Chinese Religions, 2*(1), 45–65.

Ori Tavor, PhD, is a senior lecturer in Chinese Studies and the director of the MA programme in the Department of East Asian Languages and Civilizations at the University of Pennsylvania. His research focuses on the history of Confucianism and Daoism, the relationship between religion and medicine, and ritual theory. His work has been featured in *Dao: A Journal of Comparative Philosophy*, *Body and Religion*, and the *Journal of Ritual Studies*. *Website:* ealc.sas.upenn.edu/people/dr-ori-tavor *E-mail:* oritavor@sas.upenn.edu

Processions and Masks

Facing Hardship in Ancient Europe

Matthieu Smyth

The palette of human ritual is vast: it ranges from the simplest to the most elaborate rite, from familiar gestures to intricate movements. Ordinary acts, such as gathering (a prerequisite to any ritual and already a ritual in itself), dancing, listening to a narrative, singing, playing music, sharing food and drink—a sign of human sociability par excellence—are the building blocks of ritual in the same way as are more spectacular means, such as collective trance or a transgressive celebration. As different as they may be, these ingredients represent continuity attributable to a certain unicity in human social engagement. Human interaction is thus cemented by signs of affection and marked by reciprocal exchange. Furthermore, the music, chants, dance, and all the other bodily movements create a powerful instrument, which is capable of freeing to a certain extent the participants from whatever stress they might have experienced and stored within their organism.

A procession of people donning masks or bodypainting is one of the most common human rituals. This phenomenon is known in Europe mostly through seasonal carnival celebrations: local communities stage their perception of the primal cosmic forces, notably those dwelling within the earth during winter. The procession symbolises the intrusion of vivifying chaos—beneficial or dangerous—into our world. Thus, the world of spirits, including of course the spirits of the dead, breaks into the community (Gaignebet, 1974; Lombard-Jourdan, 2005).

The most ancient forms of Carnival give us a better understanding of one of the underlying paradigms that sustained community life in early Europe—some of which were celebrated until very recently, at least within remote areas. As we reflect upon what constitutes the power of carnival rituals, we may ask ourselves if it is still possible—in a time of rampant disruption such as ours, characterised by the growing

M. Smyth (✉)
Département des Sciences Religieuses, Université de Strasbourg, Strasbourg, France
e-mail: msmyth@unistra.fr

absence of traditional rituals—to enter into fruitful dialogue with our ritual heritage (see Figs. 1, 2, 3, and 4).

Seasonal Celebrations

The origins of seasonal festivals are older than the villages of the Celts, the towns of the Roman Empire, or the centres of medieval Christianity; they are much older than the Bronze Age or even the spread of agriculture. These festivals are rooted in the cosmic order of the forests and their animals: hunting, fishing, the gathering of wild fruits, as well as the lunar cycle, which presides over all activity, form the axis of the life reflected in its rites. Their context is the world of wilderness, not yet the scene of a land brought into submission.

Carnival is the ritual parallel to the mythological Wild Hunt, day or night. For the Welsh people, the Wild Hunt was led by the beautiful psychopomp Gwynn ap Nudd, the 'white' (gwynn) god of winter, king of the 'fair folk' and master of the otherworld, the realm of Annwn. The Krampus and Perchten of Austria, the 'Wild Men' of Germany, the Tschäggättä of the Valais, the Mamuthones of Sardinia, Bulgarian Kukeri, the Masopust of Moravia and Bohemia, the Kurents of Slovenia: all the giants of animal or tree form—the sometimes comic, sometimes threatening sides—are the ritual figuration of the same myth.

Northerners of colder climates often build giant statues representing Winter, for the Carnival abolishes winter and introduces Lent. This custom is still alive in the Northern and Slavic countries: Lady Maslenitsa in Russia; Old Man Winter in Belgium; the Bööog of the Swiss Sächsilüüte, ancestor of our Bogeyman or Bugbear. After a procession is led and the Bööog is crowned, he is set on fire, a joyful scene dramatizing the seasonal transformation of the life force. This force is not dead; it is undergoing a metamorphosis in order to make its way through the trees, among the beasts, before returning to awaken the life of the community.

The Great Woodwose and the Full Moon

The masks in these seasonal celebrations represent mostly male and sometimes female savage giants, who frequently don tree-like outfits, such as the woodwoses of England. They also often embody animals, such as bears, horses, moufflons, goats or stags, that closely resemble the figure of the Wild Man. [1] These figures break into the midst of the village from the forest, wreaking havoc. The masks that were worn were often coupled with symbols either of the new moon (Harlequin or Zwarte Piet faces covered in black) or of the full moon (Pierrot Lunaire, or 'Moonstruck', faces

---

[1] Fréger (2012) amassed an astonishing collection of photographs related to the Carnival woodwose figure; see also the study by Bartra (1994).

Fig. 1 The woodwose. The four illustrations that accompany this chapter are modern renditions of ancient seasonal festivals. | © Claire Smyth

covered in white), often in parallel with representations of the female principle (these masks are also well known to us through the *Commedia dell'arte*).

The term woodwose (s. and pl.), or 'wild being(s)' refers to actual or mythical forest creatures that appear throughout medieval European artwork and literature. Although most of these hairy human-likenesses have male forms, some are depicted as wild women or families. The term 'woodwose' may also refer to forest-dwellers who were excluded from society or abandoned; the second part of the term is cognate with the German 'Waise' and Dutch 'wees', which both mean 'orphan'.

Dying and Rising Through the Seasons

Spring and summer carnivals are still held here and there. However, the majority of the carnival celebrations that survived centuries of Christianity—before surviving centuries of Modernity—took place from the full moon that occurs halfway between autumn equinox and winter solstice to the full moon halfway between winter solstice and spring equinox. These celebrations are known by their Celtic names: Samhain and Imbolc. Both are associated with the cold season: the first marks the beginnings of wintry time and the second the very first signs of spring rebirth. In Europe, especially in the north and mountainous regions, winter is the most dramatic season of the life cycle; food and warmth are scarce for humans and animals alike. Life itself seems to hibernate under the earth before being reborn as days get longer, lighter, and warmer. This transformation is often represented as a large effigy of winter's rigour or as that of food's abundance figuratively put to death at the end of winter, personifying thus the change of season. In other words, the effigy stands for a dying and rising tutelary spirit. The celebrations of All Saints, Martin, Hubert, Nicolas, Christmas, John the Evangelist, Holy Innocents, Sylvester, Epiphany, Candlemas, Antony, Brigit, Blaise, Valentine, and Shrove Tuesday, to name a few, paved the way for these animistic customs within the Christian liturgy (Walter, 2011).

The symbols that underlay the masked parade, its din and turmoil, represent the great tumult of the forest spirits invading the neat and tidy world of the living. Thus, the breath of life, notably fertility, is allowed to fulfil its task while it finds a way through our world. The nuptial Skimmington Ride (or Charivari, in which a folk parade is staged to the sound of loud and discordant music originally to celebrate marriage, or, by derivation, denigrate a marriage of which the community disapproved) represents a memory of this myth. In this respect, we could mention the various horse processions that take place around Europe in springtime (such as the Cornish 'Obby 'Oss, or the French Tarasque parade, which in fact stages a dragon). All these rituals are also reflected in the medieval belief in the Wild Hunt, which one could spot four times a year in a sudden tempest at each season (Hell, 1994; Lecouteux, 1999).

The play, laughter, banquet, masquerade, and transgression unleashed by seasonal festivals defy the threat hanging over the community at times by the forces of nature. The simple but intense joys of the feast also contribute to bringing down the boundaries between our world and the spirits' world. Here, the profane and the sacred do not split the worlds in two separate entities. Surely, the hidden world of the spirits and ours, which is visible, are not the same, but there is no transcendence to lay them out within a strict hierarchical relationship. On the contrary, they are superposed and imbricated. Full of laughter, racket, and movement, seasonal festivals do not make much room for the silence and reverence that would be inspired elsewhere by the powerless fear felt in front of an unworldly radically Other.

These features are much older than the settling of the Celtic civilisations, the Roman Empire, or medieval Christianity. The core of these festivals goes back to an animistic past, to the time of our hunter-gatherer ancestors. After all, a shaman with stag attributes was painted on the walls of the Trois-Frères cave (south-west France) during the Palaeolithic Age. [2] The forest and its animals are the cosmic background.

---

[2] See the groundbreaking study by Clottes and Lewis-Williams (1998) on prehistoric European shamanism.

Fig. 2 Spirits of the forest. The thick forests that covered Europe were home to the spirits of trees and springs commonly described in ancient European lore. | © Claire Smyth

These rituals presuppose a pattern of hunting, fishing, and berry gathering, as well as the moon cycle that presides over it. It is wilderness that provides food, not yet the domesticated world.

Life Depends on Death

Life and death are perceived and pictured through their cyclical interlocking: the life of some feeds on the death of others. All life depends on death. No birth or rebirth may take place without death. As for death, she is always pregnant of life, and thus associated with fertility—indeed a crucial aspect of life. This is a cornerstone of animism. In Celtic mythology, the Irish Dagda and the Gallic Sucellos, as well as— based upon a similar archetype—the Welsh guide of departed souls Angau (the Briton Ankou), represent this interweaving of life and death. These figures reign over the perpetuation of the cycles of life, be these daily, seasonal, parturient, or funerary. Actually, according to the animistic mindset, death and the accompanying migration of souls are not to be excessively feared.

In the face of death, as well as before the small and great stages of life, the community builds through these rituals what constitutes its very self. With the help of all, it is each one's *being there to the world* that is asserted through the celebrations staging the relationship of unity, which the community upholds in union with the cosmos. Everyone finds among the rituals some strength to keep on following his own cycle of life, convinced that such a cycle is itself inscribed within larger cycles.

Of course, there are many other rituals that nourish social bonding among human communities, such as simple greeting rituals. Therapeutic rituals take care of particular collective or individual situations; in a way, some aspects of funerary rituals belong to this category. Seasonal rituals achieve the heightening of social unity, even if they do help to overcome the individual fears that the winter phase of year's cycle begets. They involve a well-structured community whose individuals are not threatened by overwhelming stress. Rituals both reflect and enhance social unity. They are thus able to complete their task: regulating the core of each one's nervous system through social bonding. Rituals operate as a crucial tool to keep each one's life balanced and stable. Furthermore, through the music, chanting, dancing, peregrination (whether linear ritual walking or circumambulation), and all the other collective bodily movements involved, each participant is able to release the stress that may be stored within his nervous system.

Societies in Northern Europe only encountered and evolved agriculture, cities, and the resultant complex hierarchies at a relatively late stage in history compared to the Middle East, and not with the same intensity. Their fundamental rituals are far more ancient than the trifunctional Indo-European ideology that eventually prevailed. There are no castes: no kings, no sacerdotal order, no warriors, no labourers in service of the former.

Shamanic tasks belonged rather to certain women instead of men (Poly, 2006). This is, alas, the very tradition churchmen and lawyers at the junction between the Middle and Modern Ages came across when they believed to have discovered 'witches.'[3] The Western world, after the Neolithic age, did not fully succumb to the trend that drastically subordinated women to men (Poly, 2003). Similarly, communal decision-making institutions, such as the Nordic Ðing, kept on, in spite of chiefdoms raised to kingdoms. Likewise, a certain antagonism opposed the eastern despotic world governed by god-kings to the ancient Greek ideal of isonomy.

---

[3] See the foundational study by Munchembled (1979).

By contrast, the sacred meal characteristic of the Christian Church (and of Rabbinical Judaism, since both depend on the same ritual source) is wrought around bread and wine. The Christian movement was close to a purely urban reality up to the end of the fourth century. The oriental symbolism of its sacred meal, to be implemented all across the West, is rooted within viticulture and cereal agriculture. It is even linked to vegetarianism. Fish and cheese were part of the early Christian liturgical meal, but the Eucharist is also clearly tainted by the refusal of red meat peculiar to Late Antiquity: in Mediterranean asceticism—be it Hellenistic, Jewish, Christian, Gnostic—meat relates to sexual desire and to animal sacrifices that are seen as gross by the spiritualising philosophies of the time. To abstain from meat as well as sexual intercourse while fasting are typical features of Christian asceticism (McGowan, 1999).

Traditional Western festivals, on the other hand, are based upon excess. This implies consuming rich food in abundance without worrying about the morrow, in contrast with times of natural dearth and hardship. When the ground started to harden because of frost, then came the time to slaughter and preserve some of the livestock, chiefly pigs. However, offal, which would otherwise quickly deteriorate, was to be eaten on the spot. Later on, near the end of winter, when the remaining meat storage was about to run out, and before it might start to rot, was the last time to eat amply. It would soon be time to switch to the scarce food resources that were still available, notably freshwater fish. Precisely, in the Celtic and Germanic world, autumn and winter festive seasons rely on the celebratory sharing of a vast amount of animal meat, notably pork, during banquets. The boar and the hog were divine figures celebrated all across Europe. Furthermore, most of carnival's symbols are drawn from the trees and the large animals of the forest (at the time hogs were not enclosed). A depiction of the woodwose in the likeness of a tree is not uncommon. Even the community's location within the ecosystem underlying these festivals cannot be easily reconciled with the more advanced Middle Eastern agricultural background of Christianity. European animistic rituals stage thus the spirits of the forest sharing both animal- and tree-like shapes, displaying a relationship to the ecosystem where humans, to a certain extent at least, still belong to the wilderness.

Christianity, on the other hand, is dependent on agriculture. Its primary relationship with the ecosystem is domestication. Hence, the relationship of European rural communities with the ecosystem displayed during carnivals cannot be easily reconciled with the more advanced Middle Eastern background of Christianity that relies mainly on the domestication of the surrounding ecosystems. The latter is not involved in an intimate relationship with the environment but rather in a domination process.

A New Setting: Urban Carnival

As was the case in Imperial Rome and around the Western Empire with the popular Saturnalia festival, the December festival in honour of the god Saturn, in medieval towns, the animistic tradition was, if not superseded, at least partially hidden under a

Fig. 3 The Feast of the Ass. The Feast of the Ass was a medieval Christian holy day that recalled the Holy Family's flight on a donkey to Egypt. This celebration of the donkey—a creature central to early European economy—appears to have been a by-product of the Feast of Fools but was not considered as objectionable. | © Claire Smyth

new layer: ordinary townsfolk staged the subversion of the established order, notably with the humorous evocation of the powers that be, as well as their political and religious norms. Latin Fathers of the Church, such as Augustine (Sermon 189.1), complained that lay people were not ready to abandon the 'Liberties of December'. However, with time, in the Early Middle Ages, these old transgressive celebrations were given an additional Christian meaning. This new medieval paradigm gave birth to the winter Feast of Fools (Heers, 1983; Laharie, 1991; Harris, 2011)—or Feast of the Ass in some places—where Catholic liturgy was subverted for a while, with some consent of the Church (before its condemnation at the end of the Middle Ages).

Our own best testimony comes from the North of France in the first decades of the thirteenth century. The archbishop of Sens, Pierre de Corbeil himself, composed an ecclesial 'Office of the Feast of Fools' (*Sens, bibl, mun. ms lat. 46*) and gives us a close description (Villetard, 1907). This manuscript codified old practices of this church for the different rites and ceremonies to be celebrated in the cathedral in the days following Christmas. It is not easy to evaluate their exact ecclesiastical origin, but it is highly likely that this piece goes back to Carolingian times, when the Frankish Church manifested a growing taste for the *ludi scaenici*, or mystery plays, that were grafted onto certain liturgical practices (see for instance the dialogue *Quem quæris in sepulcro* at Easter). The Office was copied and adapted in many other dioceses of the North and East of France.

Clearly this type of liturgical drama supposed a whole cycle of liturgical celebrations mingling serious material and mockery. Yet it carried the Gospel message of human equality, already incarnate in the Magnificat: *deposuit potentes de sede, et exaltavit humiles* (Luke 1:52). The hero is none other than an ass, a figure from the cradle scene, clad in a bishop's cope and led in procession all the way into the cathedral choir: a mock bishop played by a choirboy is seated on the episcopal throne to celebrate mock services, including a parody of the Mass amidst other junior clerics in costumes. The inclusion of the donkey, as well as costumes designed with lunar or animal symbols, hint at the survival of animist rites in this context. Similarly, the Shrove Tuesday (*Mardi Gras*) parade, which developed during the later Middle Ages, displays the worldly and powerful in the same fashion as the Danse Macabre depicted on so many late Gothic churches (see Fig. 3). The pope, the emperor, cardinals, princes and princesses, the bishop, magistrates, rich merchants, all are equal to the humblest faithful through the personification of Death, by whom the round is conducted. In all these cases, Christianity found a way to adapt the earlier tradition. Inspired by the belief that possessions and honours, even ecclesiastical ones, are all vanities, and that at the end of the road stands each one's moral truth, the Church welcomed this desire to periodically transgress the very structures that held medieval society together, including hers.

Those structures, from the eleventh century on, had become more and more established within a new urban culture that followed the growth of the merchant towns. The face of Europe was altered. Before, only the regions close to the Mediterranean had been intensely urbanised. However, at the beginning of the second millennium, towns began to flourish in the midst of what had been the

Fig. 4 Danse Macabre (Dance of Death). The Dance of Death is an artistic medieval reminder of the fragility of life and universality of death. Typically the *memento mori* portrays a Figure of Death dancing alongside a pope, an emperor, a king, a labourer, and a child to show that no one escapes this final end. | © Claire Smyth

immense forest covering the north of the continent, while already existing cities grew considerably.

Carnival's Subversive Elements

When feudalism developed in rural Europe, it was not much different from ancient chiefdoms, and therefore did not produce complex hierarchies. However, the later medieval town shelters a class society where both the very rich and the very poor, the very powerful and the very humble, coexist with quite a few gradations in between. The brotherhoods, the guilds, the parishes, and the town magistrates now closely regulate the life of townsfolk. The Church, its clergy, its monks and its external symbols are everywhere to be seen.

All these social structures were understood as an absolute necessity. One cannot live without them since they are decisive in providing sustenance and security. They offer a powerful compensation against the 'psychosocial disintegration', which always threatens urban life, as defined by Bruce Alexander (2008) in the wake of Erikson and Erikson's (1998) theory. Social bonding remains strong. Human power has superseded the powers of nature, to which the medieval townsfolk are to some extent alien, even if the largest medieval cities are by no means comparable to modern industrial urbanisations.

Above all, some of these structures so critical to urban medieval life came with a price: the often terrifying, arbitrary power held over commoners by the town's rulers. It could be understood as an equivalent of the ambiguity—sometimes beneficial sometimes dangerous—essential to the world of spirits, which has been in some way replaced by the ruling class. Here, however, this fearful ambiguity is embodied in dignitaries who are all too real humans. They are quite far from the oneiric world, even if the greatest figures among them manage to interact with earlier mythical narratives (such as the King under the Mountain, which is based on the bear, or the savage giant sleeping under the mountain myth). Public figures shamelessly put their might on show. In order to heighten and protect the domination exerted through the rulers, the aesthetics of power must be displayed lavishly. As noted by Phillippe Ségur (2010), power is a monster that defies the laws of nature and does not shy away from being adorned with an otherworldly aura. Such human power is exerted over other humans as if it was more than human, or as if its subjects were a bit less than human. It does so while displaying its own sense of otherness, as if encountering the symbols of power were a sacred experience of the transcendent.

Carnival's subversive elements function as an exorcism against the fear human domination inevitably sparks: once or twice a year, one is allowed to mock this monster. The parade of the lesser people adorned with attire peculiar to worldly authorities makes it obvious that on a deeper level those figures are nothing. Carnival is also a time when criticisms that could not be uttered in other contexts get to be heard. The anxiety and anger begotten by every hierarchical society finds a safe outlet. Lesser people are not doomed to freeze into total helplessness and resignation. At least some kind of a figurative transgression is available to them.

Moreover, through the ritual peregrination, and the other animistic symbols still there to be found, not to mention the intense joy any festive assembly would create, the Feast of Fools or Shrove Tuesday keeps on strengthening and appeasing the community involved in these festivals. These celebrations allow for heightened social bonding. They are an opportunity for everyone to experience a greater mind–body regulation through all the collective festive bodily movements that tend to sooth stress.

What of our modern cities where social cohesion had to withstand the 'great dislocation' diagnosed by Karl Polanyi (1944)? Society, or what's left of it, cannot allow itself to deprive its own members of some minimal 'psychosocial integration'—as described by Bruce Alexander (2008). Such a vital element for us social animals cannot be overlooked if communities are to avoid collapse and their members' critical dislocation. To compensate for this threat, some rely on what remains of social bonds while some others form and join new 'tribes' (for want of a better word), which include some rather frightening religious movements, established around the pursuit of a common interest and often an unconventional lifestyle (Maffesoli, 1996).

Conclusion: Keys to Human Stability

The ultimate causes of modern general dislocation cannot be addressed by those who are enduring it, but most of its victims relentlessly challenge its effects for as long as their desire to live holds out. In the long run, many responses to these stressors may prove to be inadequate compensation. However, through individual friendly relationships, many people recreate some degree of psychosocial integration (Alexander, 2008), with rituals rooted in music, poetry, and friendly meals. The list could be broadened, for instance with the physical movements we define as sports and that are performed within a group, or with behaviours that enhance contact with what remains of wilderness, such as hiking. We witness here a primal desire to be connected to what remains of wilderness in our surroundings, to become embodied in the place where we actually stand, in the present, and above all to experience some social bonds. Human homeostasis cannot occur unless these needs are met. It is worthwhile to note that, in spite of all pressure of alienation, some basic rituals addressing these needs remain, albeit on a small scale.

Collective and symbolic non-verbal language is at the core of ritual. Human social unity finds expression through this language. Our fragmented societies no longer meet our need for ritual. The collective narratives upon which Carnival was based have long since vanished. So has the way of life through which these rituals were codified. They were once a language, but the code that this language upheld does not talk to us anymore. It is therefore difficult for us to benefit from the remaining vestiges of the rituals associated with these festivals (which, by the way, have nothing to do with the imperial rituals peculiar to totalitarian regimes). This means that we are left with only the most fundamental human rituals such as playful collective activities or friendly chats. In spite of the pressure exerted by modern economy, these rituals can provide us with a rudimentary experience through which we can feel close to our family members and intimate friends. This is where modern-day experience of the 'here and now' begins. This is how being present with our own bodies, and ultimately with the profound life cycles, begins.

This is not to say that the roots of ritual that plunge us in our distant collective selves, our ancient cultural backgrounds, are to be discarded. They also represent a vital need. Although Westerners tend to see themselves above this need, severing links with 'who one is' leads both community and individual on a hazardous path. The fate of aboriginal peoples shows this all too well.

Benefiting from ancient carnival processions and their mythical characters might be difficult, but such traditions cannot be overlooked. The experience that lies at the heart of ancient rituals, nor what begot them, is not elusive. Their wisdom, or at least some of its aspects, may therefore still be passed on to us. Through their study, we may experience ritual elements missing in our society, at least as a need that is nowadays hardly met: notably the strong social bond that ancient rituals fostered and the peaceful relationship with nature that they reflected, even when nature was perceived as potentially threatening. We may be able to build upon from there a new awareness of what was once lost.

References

Alexander, B. K. (2008). *The globalization of addiction: A study in poverty of the spirit*. Oxford University Press.

Bartra, R. (1994). *Wild men in the looking glass: The mythic origins of European otherness*. University of Michigan Press.

Clottes, J., & Lewis-Williams, D. (1998). *The shamans of prehistory: Trance and magic in the painted caves*. Harry N. Abrams.

Erikson, E. H., & Erikson, J. M. (1998). *The life cycle completed: Extended version*. W.W. Norton.

Fréger, C. (2012). *Wilder Mann: La figure du sauvage*. Thames & Hudson.

Gaignebet, C. (1974). *Le Carnaval: Essai de mythologie populaire*. Payot.

Harris, M. (2011). *Sacred folly: A new history of the feast of fools*. Cornell University Press.

Heers, J. (1983). *Fêtes des fous et carnavals*. Arthème Fayard.

Hell, B. (1994). *Le Sang noir: Chasse, forêt et mythe de l'homme sauvage en Europe*. Flammarion.

Laharie, M. (1991). *La Folie au moyen âge (xie – xiiie siècles)*. Le Léopard d'Or.

Lecouteux, C. (1999). *Chasses fantastiques et cohortes de la nuit au Moyen Âge*. Imago.

Lombard-Jourdan, A. (2005). *Aux origines du carnaval, un dieu gaulois ancêtre des rois de France*. Odile Jacob.

Maffesoli, M. (1996). *The time of the tribes: The decline of individualism in mass society*. Sage.

McGowan, A. (1999). *Ascetic eucharists: Food and drink in early Christian meals*. Oxford University Press.

Munchembled, R. (1979). *La Sorcière au village (xve – xviiie siècles)*. Gallimard.

Polanyi, K. (1944). *The great transformation*. Farrar & Rinehart.

Poly, J.-P. (2003). *Le Chemin des amours barbares: Genèse médiévale de la sexualité européenne*. Librairie Académique Perrin.

Poly, J.-P. (2006). Am Stram Gram: La chevauchée des chamans. In *L'Histoire*, numéro spécial *Dieu au Moyen Âge*, pp. 60–63.

Ségur, P. (2010). *Le Pouvoir monstrueux*. Buchet-Chastel.

Villetard, H. (1907). *Office de Pierre de Corbeil: (Office de la circoncision) improprement appelé 'Office des fous'*. Alphonse Picard & fils.

Walter, P. (2011). *Mythologie chrétienne. Fêtes, rites et mythes du Moyen Âge*. Imago.

Matthieu Smyth, PhD, is a ritual anthropologist at the University of Strasbourg and trained as a Somatic Experiencing® practitioner (SEP). He is the father of three children, an avid alpinist, and the author of *La Liturgie oubliée* (2003, Le Cerf) and *Ante Altaria* (2007, Le Cerf). Matthieu lives in Besançon, France. *Website:* rituelprimal.com *E-mail:* msmyth@unistra.fr

Part II
The Role of Ritual in Healing Trauma

At the Sharp End of Medical Care

Healing and Reconnecting Through Ritual

Robin Karr-Morse, Juan Carlos Garaizabal, and Jeltje Gordon-Lennox

Until 1986, newborn babies in the United States and the United Kingdom were routinely subjected to major surgery to repair malfunctioning hearts, lungs, and kidneys—without any anaesthesia at all. [1] Many died from pain and shock (Chamberlain, 1995). The word 'trauma' tends to evoke images of sudden disasters such as a car crash, a terrorist attack, war-related events, or natural catastrophes like an earthquake or a flood. It is rarely associated with sophisticated modern medical procedures. Yet there is more and more evidence that, in even the best hospitals and clinics around the world, children and adults alike are regularly traumatised during medical procedures.

Trauma is a fact of life. We all have experienced or will experience terrifying events, but not all of us bear the long-term effects of trauma. Psychologist Peter A. Levine holds that our bodies and minds are designed to heal intense and extreme experiences. People become traumatised when they lose their capacity to self-regulate arousal, orient, be present, and flow with life (Levine, 2010). One way to maintain that capacity, suggests Levine, is through direct participation in the creation of our own transformational experiences through ritual (2005).

---

[1] As barbaric as that may seem to us today, it was long held that the nervous systems of newborns were too immature to feel pain. In addition, until recently, it was extremely difficult to gauge how much anaesthetic a newborn could tolerate.

R. Karr-Morse
Psychotherapist, Portland, OR, USA
e-mail: robin@theparentinginstitute.com

J. C. Garaizabal (✉)
Voice Movement Therapist – Vocal Coach, Bioenergetic Analysis Therapist, Certified TRE Provider, Bilbao, Vizcaya, Spain
e-mail: info@juancarlosgaraizabal.com

J. Gordon-Lennox
Psychotherapist ASP, Geneva, Switzerland
e-mail: jeltje@gordon-lennox.ch

© The Author(s), under exclusive license to Springer Nature Switzerland AG 2022
J. Gordon-Lennox (ed.), *Coping Rituals in Fearful Times*,
https://doi.org/10.1007/978-3-030-81534-9_6

Ritual in Medical Care

The word 'ritual' evokes various images, many of which might seem out of place in an article about medical care. Yet the first usage we find of the term in this context dates back one hundred years:

> [The success of an operation in surgery depends] not only on the skill but also upon the care exercised by the surgeon in the ritual of the operation. The 'ritualist' must not be a man unduly concerned with fixed forms and ceremonies, with carrying out rigidly prescribed ordinances.... The ritual of an operation commences before, and sometimes long before the incision is made and may continue for a long period after the wound is healed. In the craft of surgery the master word is simplicity. (Moynihan, 1920, pp. 27–28)

The Body Bears the Burden

An adult with a fully developed nervous system—and hopefully a wealth of healthy experiences with kind caregivers—is far better equipped to calm the strong emotions that arise and self-regulate during a medical procedure than a newborn baby or child. Infants—whose nervous systems are as yet unmyelinated [2]—depend on adults with mature nervous systems to calm them and teach them how to self-regulate. A single overpowering event such as the death of a parent can over stimulate the child's undeveloped system and overwhelm their ability to cope. But, so can an accumulation of 'lesser' stressful experiences such as a mother's highly stressful pregnancy; prenatal alcohol consumption; induced labour; premature or caesarean birth; early surgical procedures; chronic pain; separation from the primary attachment figure through death or hospitalisation; inappropriate, inadequate, or unpredictable child care; foster care or adoption; desertion or neglect; maternal depression; domestic violence; divorce; or insensitive custody arrangements. Most of the above-mentioned events are unintentional and result from ignorance; none of these experiences alone is likely to compromise a child's health. An accrual of these situations, however, can cause severe damage to young nervous systems.

Traumatic stress responses during the earliest periods of life—when the child has not yet acquired language—are particularly damaging because the brain records them primarily as somatic, rather than conscious, experiences. Early life trauma in children is compounded by overstimulation (especially from pain and neglect) and/or under-stimulation (resulting most profoundly from separation from the mother or main caretaker). 'Feeling' memories are registered in procedural (or unconscious) memory, rather than as a declarative or verbal memory. Early

---

[2] Myelin is a fatty substance formed in both the central nervous systems (CNS) and the peripheral nervous systems (PNS). It insulates nerve cell axons to increase the speed at which information travels among nerve cell bodies (CNS) and from nerve cell bodies to muscles (PNS). In humans, most neurons are unmyelinated at birth. During infancy, myelination progresses rapidly. This process corresponds with the child's development of cognitive and motor skills, as well as acquiring memory and reacting to the sensation of pain.

medical events experienced as trauma are stored as speechless sensations, leaving the child—and later the adult—with no memories to explore or words to explain what happened. Juan Carlos Garaizabal tells of the circumstances that led to his early experience with terror and helplessness and how this affected his development (see Case study 1a).

Case study 1a

As a baby, I never crawled but pushed myself across the floor with my feet while leaning on my hands and arms. When I was three years old I was diagnosed with *genu valgum*,[3] commonly known as 'knock-knee', and admitted to the local children's hospital for corrective orthopaedic surgery (see Figs. 1 and 2). During the first two weeks after admission, children were allowed no visits from family members. I found myself alone in a ward ringed with small beds along with a lot of other children who cried out for comfort night and day. I barricaded myself behind a wall of silence and muscular tension, lost my appetite and quite bit of weight.

I remember waking up one day totally disoriented and unable to move. It was a terrifying experience. When the effects of the anaesthesia finally wore off I realised that I was immobilised by a plaster cast from waist to toes; two holes in more or less the right places allowed me to relieve myself. For three months, I was bedridden. Liquid intake was limited to prevent bed-wetting. I felt physically exposed and vulnerable.[4] Each time I tried to stretch my hips and legs, my body was seared with excruciating pain. The mere sight of a doctor's white coat was enough to make me scream with fear. As an adult, my body still associates any kind or level of physical immobility with intense feelings of confusion, helplessness, madness, and fear of death.

Later, when I was old enough to understand, my mother told me about the succession of medical and bureaucratic errors that had overwhelmed her. Two inexperienced doctors performed an orthopaedic operation on me that consisted of cutting the bone below the knees to give another axis of symmetry to the legs. They should never have operated both legs at the same time. Moreover, since this public hospital offered minimal post-surgical physical therapy, they made

(continued)

---

[3] *Genu valgum* describes a usually benign variation in knee position that may be caused by injury or infection of the knee or leg, obesity, or early vitamin deficiency. The condition does not appear to be genetic but it may run in families. It is often first observed in children between the ages of two and five. Knock-kneed children also tend to have flat arches and feet that point inward, which further complicates their learning to walk. Depending on the cause of the condition, *genu valgum* may well correct itself as the child grows. In the past, surgery was often prescribed at a young age to correct the angle of the knee. The treatment of choice for *genu valgum* today is less radical: massage of leg muscles combined with gentler therapies, orthotics, vitamins, and, in the case of obesity, weight loss.

[4] Gordon-Lennox has clients whose recollections of medical interventions during childhood on or involving genitalia resembled the reactions (shame, fear) of victims of childhood sexual abuse.

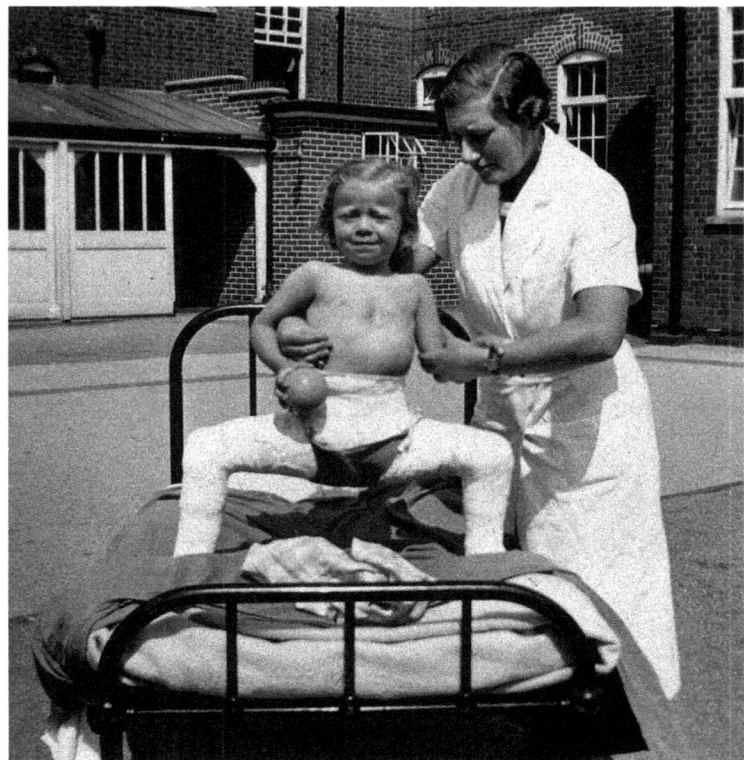

Fig. 1 Girl treated for *genu valgum*. Nurse supports the child who wears a plaster cast that covers her from the waist down. | © Wellcome Collection CC BY

my mother responsible for making me perform the extremely painful exercises. Despite our best efforts, well into adolescence I continued to fall, often and hard. But, my knees were not the only part of me to bear scars. The double role my mother was forced to assume put undue strain on our relationship at all levels, physical, emotional, and psychological.

Subsequently, we learned that I should never have been subjected to these operations at all. In cases like mine, where the bones are not deformed, the problem usually resolves itself as the child grows. Gentle massage treatment to stretch and lengthen leg muscles should have sufficed to straighten my legs.

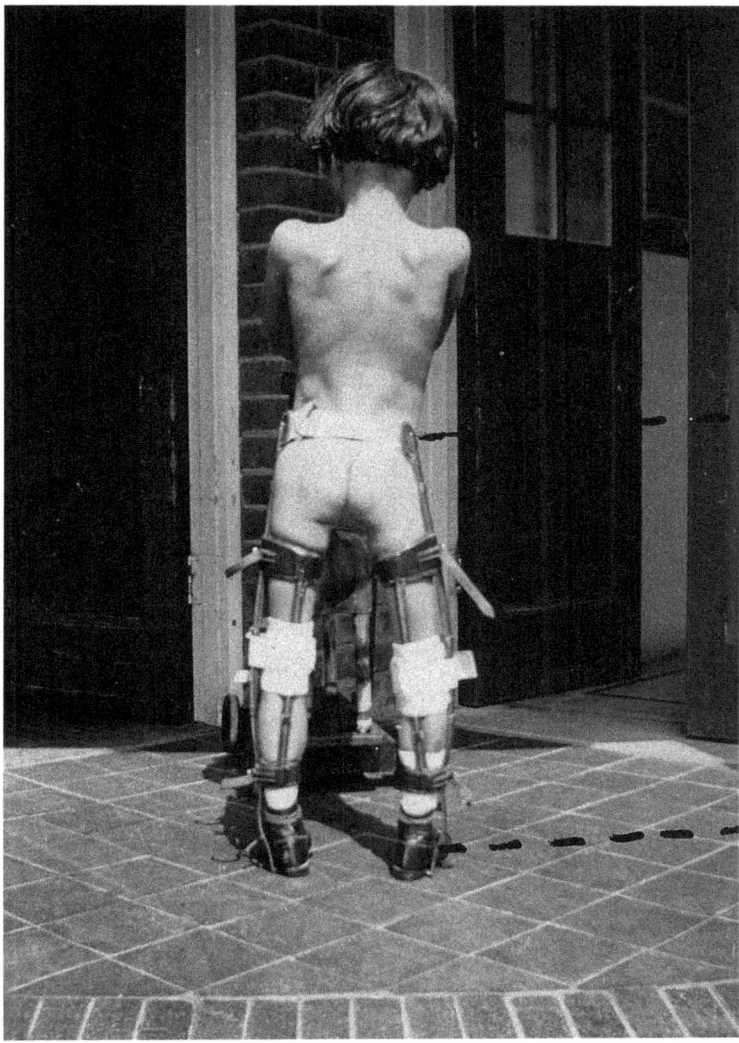

Fig. 2 A child in callipers. Correction of *genu valgum* with leg braces. | © Wellcome Collection CC BY

At the Sharp End of Care: The Patient

Why do children like Garaizabal develop dissociative stress disorders?[5] Recent advances in biology validate what many of us have known for a long time. What happens to our bodies affects our emotions, just as what happens to us emotionally affects us physically. Traumatic experiences influence how the child deals with the world and even the choices she or he makes as an adult. As the child's body grows into that of an adult, it bears the burden of trauma.

Positive emotions such as love, gratitude, feelings of connection and attachment to people, and a sense of achievement or accomplishment are healthy and healing experiences that help mitigate disease. But the opposite is also true. Chronic negative emotional states—especially fear, but also early chronic experiences of frustration, rage, shame, and grief—can activate whatever genetic tendencies we may have for disease. For some, that may mean heart disease, for others cancer, diabetes, arthritis, depression, or fibromyalgia.

Two new factors emerging from research in human physiology help us understand more clearly how chronic negative emotions and trauma increase risk of disease. The first factor involves fresh insights on health from the field of genetics. More precisely, it reveals the role of epigenetics on health. The second factor involves how protracted threat response—commonly referred to as fight, flight, freeze (immobilisation)—impacts our bodies.[6]

Role of Genetics in the Risk of Disease

Let's begin with genetics because that is where most of us start. This includes many doctors whose first reaction to a diagnosis of, say high blood pressure, high cholesterol, or arthritis is: 'Well, this is likely to be genetic since your Dad and Grandpa had the same problem.' In fact, genes alone may not be to blame for many of the problems we just noted. A closer look at these ailments shows that many have been regarded as 'genetic' because that is how a child's body learns to respond to stress, threat, and tension in the family circle, often long before she or he begins school.

Recent findings in genetics expose some surprising new facts, such as the possibility that qualities acquired from parents' experience can be transmitted to

---

[5] Neurologist Robert Scaer holds that children, who have no capacity to fight or flee, almost inevitably suffer from a later dissociative disorder, and not from the symptoms of post-traumatic stress disorder (PTSD). As a result, they seldom exhibit symptoms of PTSD, but rather fall into the cognitive state of sensory processing disorder (SPD), analogous to ADHD, or DESNOS (disorders of extreme stress, not otherwise specified) (Scaer personal communication with Gordon-Lennox, 2016; Scaer, 2017).

[6] Polyvagal theory identifies fight and flight reactions with the sympathetic state and immobilisation with the dorsal vagal state (Porges, 2011).

future offspring (Begley, 2009). Along with DNA, [7] which remains stable, our cells inherit dynamic chemical processes that surround genes and give them instructions.

If DNA contains the blueprints or codes that form the building blocks of life, the non-coding proteins that circulate around DNA—much like microscopic satellites that rapidly read and respond to their surroundings—regulate how DNA functions by detecting and responding to environmental influences. These non-coding proteins react to a host of variables including viral infections, hormones, diet, and toxins like alcohol, prescription and over-the-counter drugs, plastic by-products and other, often unrecognised, toxins in the environment. Non-coding proteins also determine which genes are expressed and which are repressed. The unlimited number of possible combinations makes predictions based solely on DNA impracticable. These processes, called 'epigenetics' [8]—literally 'above the genome' [9]—occur around DNA but do not change its basic order.

Epigenetically Acquired Protection

Scientists suspect that epigenetics account for up to 70 per cent of the risk for a given disease. Even more astonishing is the fact that we are not only talking about our own experiences but also those of our parents or even grandparents. The number of genetic variables that can influence the outcome of a disease does indeed appear limitless. Nature is full of examples of organisms that evolve or are shaped by their external environment.

Think of the 'water flea'. In spite of its name, it is not an insect but a tiny crustacean often found in freshwater streams and ponds. What's remarkable about the water flea is that when you examine a group of them—all with identical DNA—you see that some of them are bareheaded and others sport a spiny sort of protection, a bit like a helmet, which safeguards them against predators (see Figs. 3 and 4). Why are some protected and others not? The answer is that if the mother water flea has a bout with a predator, her babies are born with thorny headgear (Petrusek et al., 2009). A mother whose life has not been threatened produces young without thorns on their heads. The mother's experience with danger literally armours her offspring

---

[7] DNA is sometimes referred to as the blueprint of life because it contains the instructions needed for every living organism to grow, develop, survive, and reproduce. DNA does this by controlling protein synthesis. Proteins do most of the work in cells, and are the basic unit of structure and function in the cells of organisms.

[8] The term 'epigenetics', a blend of 'epigenesis' and 'genetics', was coined by C.H. Waddington in 1942. As the distinctions between genetic and epigenetic are increasingly blurred, current use of the term refers to effects that do not involve DNA base sequence changes but only the chemical modifications of DNA is rapidly becoming obsolete. Epigenetics is now commonly used more broadly to mean hereditary influences arising from environmental effects in the course of development (Ho, 2017).

[9] The term 'genome' refers to the complete genetic information (either DNA or, in some viruses, RNA) of an organism.

Fig. 3 Daughters of a threatened water flea sport thorny headgear. These specimens of *Daphnia atkinsoni* each sport protective headgear or a 'crown of thorns' (see spike ridges along right side of head in Fig. 4). This defensive trait is induced in offspring only when the mother senses chemical cues released by one of its main predators, the tadpole shrimp *Triops cancriformis* (Petrusek et al., 2009). | Permission granted

Fig. 4 Close-up of a water flea daughter's thorny head-gear. | Permission granted

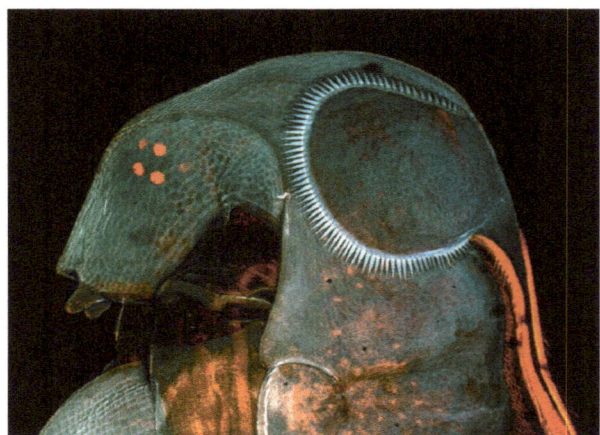

with protective headgear! We can say that these babies with headgear have been epigenetically prepped to survive an attack.

The snowshoe hare is another example that contradicts our view of genes as an impermeable fortress. When there is proliferation in the predator population, the mother hare secretes stress hormones that have both a contraceptive and protective effect. There is a decline in birth rate and a rise in the reactivity of the nervous systems of the babies produced. Like their mothers, the young are hypersensitive and extra vigilant. Once again we see nature's design for the survival of a species. (Imagine how this might apply to a human mother who has to deal with predators in her environment!) Research on children and grandchildren of people who survived the Holocaust confirm this hypervigilance in their descendants as well as increased vulnerability to ensuing trauma (Yehuda et al., 2005).

Threat, Trauma, and Disease

The second physiological factor that is key to understanding the trauma/disease connection is the effect on the body of the sympathetic state often called 'fight/flight', more specifically the impact of trauma on the immune and endocrine systems. We have known for some time that 'stress' can have a negative effect on health. But, prolonged toxic stress is more likely to pave the way to chronic disease when it is experienced early in life. Here's how it works: mobilisation through fight or flight when confronted with a stimulus perceived as an imminent threat to survival is quite well known. What happens—as in the case of infants or very young children—when neither fighting nor fleeing are realistic options?

Helplessness and Terror During Immobilisation

Feelings of total helplessness and terror in the face of real or perceived danger, such as that described by Garaizabal, will trigger a third response, which is observed by biologists as an external state of 'immobilisation', and considered by psychologists as an internal 'dissociative' state (Levy, 1944).

Children are particularly susceptible to immobilisation when stressed or threatened. Tremendous energy is summoned as an adaptive response to threat and stress. When this energy cannot be released in the sympathetic state of fight or flight, it overwhelms the brain, much like an electrical appliance receiving too strong a current. Garaizabal describes the feeling as tension and silence that builds up to block out whatever is too scary to take in. In fact, parts of the brain go offline, leaving the victim speechless. This protective response of the parasympathetic system to threat puts the victim into a dorsal vagal state. Not only does the mind go blank, but the body goes into metabolic shutdown; endorphins—the body's own calming and painkilling agent—are released to provoke a numbing dissociative state that essentially prepares the body for death.

Scared Sick

What might have been adaptive for a child facing an episodic event—that is, dissociating or immobilising during an event beyond their control—becomes maladaptive when this response remains 'turned on' even in environments that are objectively safe. Children may well carry this distressingly maladaptive and even life-threatening response, known as psychophysiological trauma, on into adulthood. What happens to traumatised children like Garaizabal when they grow up?

Neurobiological research suggests continuity in the expression of stress coping deficits among abused and neglected children over the course of their entire life span. Most often trauma in the developing child is experienced as fragmentation or disembodiment that affects relationships and health (Scaer, 2014 [2001], Scaer, 2005, 2012; van der Kolk, 2014). Psychiatrist Bessel van der Kolk (2014) explains that when the connections within the brain as well as those between the mind and the body are disrupted through trauma we become trapped in the emotions and feelings of the past. Inordinate amounts of energy are expended just to keep these sensations under control—usually at the expense of concentration, the ability to memorise, and the ability to simply pay attention to what is happening around us. Neurologist Robert Scaer describes this frightening experience an aberration of memory (2014 [2001]). The inability to live fully in the present impedes adequate preparation for the future, which in turn wreaks havoc on health and social relationships such as marriages, families, and friendships (Scaer, 2005, 2012).

Developmental Trauma

Accumulated stress has been linked to inflammatory diseases and syndromes closely associated with the freeze response, including fibromyalgia, chronic fatigue syndrome, irritable bowel syndrome (IBS), piriformis syndrome (Scaer, 2014 [2001]), gastroesophageal reflux disease (GERD), HPA axis dysfunction, and mitral valve prolapse. What each of these maladies has in common is a cyclical dysregulation of physical functions for which there is no discernible anatomical abnormality (Morris et al., 2019).[10] Medical personnel often treat these patients in confusing and inefficient ways. Moreover, the patient, who already feels helpless, may be blamed for that fact that their clinical state cannot be managed by traditional medical treatments.

Scaer treated many people who moved from the helplessness of the freeze response to empowerment. In his view, somatic methods that incorporate attunement and bi-hemispherical sensory stimulation are important for treating dissociation and trauma. But, he warns, effectiveness is dependent on the social engagement[11] provided by the therapist. He posits ritual practice in a group

---

[10] There is strong evidence that childhood social economic deprivation (SED) and/or adverse childhood experiences (ACE) are associated with the development of adult HPA axis dysfunction and neuropsychiatric, neurodegenerative, and autoimmune illnesses. Important contributors to the allostatic load experience during childhood include mitochondrial dysfunction, and nitrosative and oxidative stress and inflammation, which in turn affect the regulation of HPA axis activity, including via epigenetic factors (Scaer, 2014 [2001]).

[11] In this context, 'attunement' refers to the process of being in tune with another human being or group. It implies an understanding of what is needed or wanted by that person or group (Scaer, 2017).

setting as the essential ingredient for the resolution and healing of life trauma. 'The burgeoning growth of "somatic psychology" regarding trauma may well ensure this connection' (2017, p. 63).

The Healing Power of the Voice

The immobilisation, or dorsal vagal, state is basically a survival strategy that we share with early vertebrates. Wild mammals may 'feign death' in an attempt to escape a predator. When the danger is past, they appear to shake it off and be none the worse for the experience. Caged and domesticated animals and humans retain this amazing capacity for immobilisation; they also share a limited ability to just shrug off the intense feelings of helplessness and get on with life.

The initially protective freeze state is frequently accompanied by weak muscle tone as well as physiological changes, such as a sudden drop in blood pressure, slowed heart rate, and shallow breathing. This physical shutdown creates social distance: as muscle tone slackens eyelids droop, the voice loses inflection, positive facial expressions disappear, awareness of the sound of human voices becomes less acute, and sensitivity to others' social engagement behaviours decreases (Porges, 2011). One of the dreadful long-lasting consequences of surviving an immobilisation with fear state involves having it re-triggered, unbidden, by a particular movement, smell, taste, touch, or sound.

Remarkably, on the flip side, these same elements can be transformed into healing strategies. As humans intentionally modulate their facial expressions (laughing), the intonation of their voice (engaged storytelling), their breathing patterns (singing, playing a wind instrument, practising pranayama yoga), or even shift their posture (yoga, Somatic Experiencing, Rolfing, the Alexander Technique, Feldenkrais Method), they can also change their physiology and move towards health. The benefits of these ritualised (intentional) movements occur primarily through manipulating the function of the myelinated vagus to the heart (cf. Porges' chapter in Part I). Garaizabal's experience is consistent with these findings (see Case study 1b).

> **Case study 1b**
> At some point while I was in hospital there must have been a man who showed me kindness. I remember only a presence and the sound of a reassuring masculine voice at my side. Later, as a youngster, I noticed that when I sang my fears faded and a frozenness deep inside me began to melt; I felt increasing relief from my internal tensions. Just as each footstep was a painful struggle for me, each note of music brought me comfort and joy. Hearing my own voice sing made me feel safe. Singing was my first and most important step on a long path of natural healing.

(continued)

Fig. 5 A family in a hospital waiting room. Critical illness and injury in childhood is a source of distress not just for the child but also for the entire family. | © rawpixel

During long walks in the mountains, singing along with other hikers became an informal joyous ritual that filled me with vitality and pleasure—and it enhanced my feeling of being part of the group. Using my voice in song changed my relationship to others because I could feel that they saw me in a more positive light. Singing also changed how I saw myself. My voice created a physical space where I could be myself when I was in a group of people. Being able to let myself go in song gave me the sense of freedom I needed to explore life with curiosity and delight. As a teenager, I loved playing records and singing along with the lead singer. Intuitively, I chose songs with lyrics, melodies, and instrumental arrangements that expressed my emotions, my world, and my identity. Singing along with these artists reinforced my positive image of myself.

Even today my voice serves as a frontier that both unites me with others, and differentiates me from them. Now, as then, singing is one of the creative and recreative rituals that gives my life a playful structure through which I can release emotions and physical tensions, and experience personal and interpersonal transformation.

At the Sharp End of Care: The Careworn

Patients are not the only ones to feel the sharp end of care. Critical illness and injury take a toll on caregivers too (see Fig. 5). Seeing someone you love suffer terribly, knowing that you are utterly helpless to relieve their pain, can provoke exactly the same traumatic effects as those experienced by the patient. In addition, the caregiver who finds themselves responsible for a sick family member also has the difficult task of liaising with ever-changing medical personnel, administrators, or even lawyers (e.g. to obtain therapy, medicaments, support, insurance, and reimbursement).

Sooner or later, critical and chronic conditions have devastating effects on parents, siblings, and marital cohesion. Permanent impact of the illness of a sibling is frequently underestimated. Many studies reveal that the basic needs of family members (i.e. rest, nutrition, communication) often go unmet. [12] Caregivers may be unable to cope because they lack sufficient means or maturity, or suffer from addiction, or physical or mental illness.

The diversification of family structures and geographical mobility adds new dimensions to the impact of critical illness and injury (Shudy et al., 2006). Job transfers, political and social upheaval, or immigration may put families thousands of miles away from their traditional support systems and spiritual resources. The face of the careworn may belong to a single parent, a grandparent, aunts, uncles, and older siblings who act as primary caregivers. Members of 'blended' families created by new marriages and/or domestic partnerships, multigenerational families, or foster caregivers may unexpectedly find themselves with responsibility for the care of the sick person, but without legal rights to information or the decision-making process.

At the Sharp End of Care: The Professional

Medical personnel are also susceptible to trauma (see Fig. 6). They too have to deal with patient suffering as well as economic, technological, ethical, and legal challenges. One doctor recounts how some forty years ago he performed fourteen endoscopies a day assisted by a single nurse. Now, in spite of his greater experience and assistance by two nurses, he struggles to complete ten procedures a day. Preparation for the procedure resembles a ritual composed of stilted language, performed in a ceremonious fashion by an 'officiant' (nurse, physician). 'The ritual is not part of the endoscopy but it serves to reduce fear, prevent disruption, and maintain order' for the patient and healthcare providers (Sonnenberg, 2017, p. E628).

Underlying this new concern for caution during procedures is another rarely acknowledged sharp edge to care: medical error. Although most care providers have either been directly involved in an adverse medical event or witnessed a colleague make a medical error, as was the case with Garaizabal's unnecessary

---

[12] A higher percentage of mothers show symptoms of psychiatric disorder and/or PTSD than fathers at admission and after discharge of their child from paediatric intensive care units (PICU) (Shears et al., 2005). Mothers of critically ill children who were diagnosed with an illness or injury that would have a chronic component had increased psychological distress and decreased well-being as compared with mothers of those children with time-limited illness or injury (Tomlinson et al., 1995). Moreover, parents of children with life-threatening illnesses are more often likely to receive support from health care personnel than parents of children with chronic illnesses (Katz, 2002).

Fig. 6 The care team. First aid workers, doctors, nurses, psychotherapists, humanitarian volunteers, and others who work on the frontline of disease and disaster are vulnerable to trauma. | © Chanikarn Thongsupa/rawpixel

operations, these situations are often shrouded in secrecy. The patient is nearly always left in the dark about what happened and why (see Case study 2).

Case Study 2

Medically Induced Trauma

Linda Kenney was admitted to hospital for what was supposed to be yet another intervention on her malformed ankle. She came out ten days later after having survived full cardiac arrest and a host of complications resulting from a medical accident that no one was willing to admit happened. She was told only that she had experienced an allergic reaction to the anaesthetic and was fortunate to progress to full physical recovery.

Risk or Trauma Management: A True Choice?

From the risk management perspective, everything had gone to plan. The patient had been successfully isolated, treated, and released from the hospital without further incident, or bad publicity. Until Fredrick van Pelt, her anaesthesiologist, stepped forward. Van Pelt recounts how he and the team focused on the resuscitation by keeping their emotions on hold. In the aftermath, the impact of the accident began to sink in.

A Compelling Urge to Do Something
Van Pelt's compassion for the patient compelled him to convey his apologies to her and her family for his role in the situation.

I felt personally responsible for what had happened and compelled to communicate with the family. I thought I would be able to provide a factual account of the event to the husband but to my shock, the husband came at me with full emotional and physical force; fortunately the orthopaedic surgeon intercepted him. I was now forced to confront my own emotional distress and I realised my complete lack of training in how to manage this situation.

Without informing the hospital, I wrote [the patient] a letter in which I acknowledged the emotional impact that this event had had on her family, as well as on myself. I apologised for my responsibility for having initiated this sequence of events, and I invited the patient to open communication if and when she was willing and interested. (van Pelt, 2008, p. 249)

On her side, Kenney recalls:

When I realised what had happened, my initial reaction was gratitude just to be alive. My family, however, was quite traumatised. They had experienced a side of the event that I could not understand. For several weeks my husband cried every time he looked at me. My children who were 13, 12, and 3 at the time were each handling it in their own way.

One week after I arrived home, I received a letter from Dr van Pelt. He acknowledged the impact this must have on me and my family and informed me of how deeply this had affected him as well. He apologised for what had happened saying that he believed in open and honest communication. He made it clear that he would be available if I wanted to speak to him further. I then called Dr van Pelt so I could let him know that I didn't blame him and that I believed this was a truly unanticipated outcome. This conversation was extremely healing for me and enabled me to move on. (Kenney, n.d.)

Restoring Broken Connections
Six months after the accident, the doctor and the patient sat down to discuss what had happened. Their conversation ended that day with the patient

(continued)

offering the doctor forgiveness. 'It was one of the most powerful and uplifting experiences in my life and served to set my life in an entirely new direction,' remembers van Pelt. 'In an instant, the burden that I had been carrying on my chest had lifted and I was free' (2010).

A straightforward discussion freed them both in an extraordinary way to pursue their respective personal and professional goals. Kenney dared to return to hospital for another operation, and accompany her family in their fear of the consequences of her decision. Even though van Pelt's colleagues perceived it as disloyalty to the care team, he assumed his blatant disregard for his hospital's risk management procedures.

Van Pelt and Kenney's concern about the impact of medically induced trauma on his professional care team and her family compelled them to act together to change the system that had failed them both. 'I began to realise that this was bigger than me,' recalls Linda. 'I experienced an incredible sense of responsibility because I was one of the lucky ones to have survived. This hole in the system needed to be filled' (Kenney, n.d.).

A Tragedy Transformed by Transparency and Apology

When two people at opposite ends of the sharp edge of medical care join forces to do something, they can change awareness and support around adverse medical events. The profound need of one doctor and one patient to effect change resulted in the creation of a not-for-profit foundation, Medically Induced Trauma Support Services (MITSS). Unexpectedly, it also catalysed the creation of a hospital-based peer support network for care providers at van Pelt's hospital. 'To our great surprise,' van Pelt says, 'these two initiatives that had blossomed out of a tragedy transformed by transparency and apology were all of a sudden propelled to national and international attention. We had awakened a sleeping giant in healthcare quality and safety that could no longer be ignored' (2010).

Conclusion

Patients, their entourage and medical professionals may all find themselves at the sharp end of medical care. The rituals surrounding medical procedures can serve to reduce fear, prevent disruption, and maintain order for all three groups.

In the case of Garaizabal, thanks to a vague memory of a reassuring voice in hospital, he spontaneously began using his own voice in song to calm himself. Once he had experienced how his voice could naturally connect him to others, he built on that sensation with creative rituals around song that enhanced self- and co-regulation as the child moved towards social integration.

As an adult, Garaizabal was instinctively attracted to Alfred Wolfsohn's (1896–1962) [13] work. He travelled from his home in Bilbao to London in order to learn Voice Movement Therapy. After that, Garaizabal also trained in various other body-based therapies, including Trauma Releasing Exercises (TRE), all of which he used to cure himself and then to accompany his clients on their path to healing. Garaizabal's mother cried with her son when she read a draft version of this chapter: finally her suffering was recognised.

The efficient use of simple intentional rituals–such as a face-to-face encounter where one says 'I'm sorry. . .' and the other replies 'I forgive you'—can promote a physiological state of safety in which the outcomes of trauma resolution are optimised. As van Pelt and Kenney responded in an unexpected manner to their compelling need to do something to palliate their dis-ease, the rituals they practised safely channelled the strong emotions that had flooded them in the aftermath of the adverse medical event and promoted self- and co-personal regulation. Later, by daring to share their story and what they had learned from this process, the patient and the doctor brought healing and social regulation to their respective support communities. By joining forces, the two founders continue to touch people they may never meet in ways they could scarcely have imagined during their first conversation. This how an adverse medical event remarkably transformed into a not-for-profit foundation that continues to spread healing and restore broken connections among those at the sharp end of care.

Coming back to Moynihan, we see that his one-hundred-year-old wisdom about the role of ritual in medical interventions still applies today. The rituals of medical intervention do indeed begin before, sometimes long before, any medical gesture is made. The care transmitted by intentional gestures and words can reduce fear and go a long way towards preventing and healing medically induced trauma among patients and healthcare providers. Practising simple rituals that involve transparency and compassion—even long after the physical wounds are healed—promotes healing and connection at all levels of medical care.

References

Begley, S. (2009). From 'the sins of the fathers' to the virtues of the mothers: Lamarak won't go away. *Newsweek*, 3 February. Accessed August 1, 2020, from www.newsweek.com/sins-fathers-virtues-mothers-lamarck-wont-go-away-221830

Chamberlain, D. (1995). What babies are teaching us about violence. *Pre- and Perinatal Psychology Journal, 10*(2), 5175.

Ho, M. W. (2017). *The meaning of life and the universe: Transforming*. World Scientific.

---

[13] Wolfsohn was a German singing teacher who was diagnosed with shell shock, after serving as a stretcher bearer in the trenches during World War I. Although Wolfsohn's persistent auditory hallucinations of screaming soldiers did not respond to the classic treatments of the time, he cured himself by vocalising extreme sounds, bringing about what he described as a combination of catharsis and exorcism.

Katz, S. (2002). When the child's illness is life threatening: Impact on the parents. *Pediatric Nursing, 28*, 453–463.

Kenney, L. (n.d.). *Linda's story*. Medically Induced Trauma Support Services (MITSS). Accessed August 1, 2020, from https://betsylehmancenterma.gov/news/five-questions-with-linda-k-kenney

Levine, P. A. (2005). Foreword. In M. Picucci (Ed.), *Ritual as resource: Energy for vibrant living* (pp. xvii–xix). North Atlantic Books.

Levine, P. A. (2010). *In an unspoken voice: How the body releases trauma and restores goodness*. North Atlantic Books.

Levy, D. M. (1944). On the problem of movement restraints. *American Journal of Orthopsychiatry, 14*, 644.

Morris, G., Berk, M., Maes, M., Carvalho, A. F., & Puri, B. K. (2019). Socioeconomic deprivation, adverse childhood experiences and medical disorders in adulthood: Mechanisms and associations. *Molecular Neurobiology, 56*(8), 5866–5890. https://doi.org/10.1007/s12035-019-1498-1

Moynihan, B. G. A. (1920). The ritual of a surgical operation: Remarks made at the opening of a discussion at the first meeting of the British Association of Surgeons, held at the Royal College of Surgeons. *British Journal of Surgery, 8*, 27–35. https://doi.org/10.1002/bjs.1800082906

Petrusek, A., Tollrian, R., Schwenk, K., Haas, A., & Laforsch, C. (2009). A 'crown of thorns' is an inducible defense that protects Daphnia against an ancient predator. *PNAS, 106*(7), 2248–2252.

Porges, S. W. (2011). *The polyvagal theory: Neurophysiologial foundations of emotions, attachment, communication, and self-regulation*. W.W. Norton & Company.

Scaer, R. C. (2005). *The trauma spectrum: Hidden wounds and human resiliency*. W.W. Norton & Company.

Scaer, R. C. (2012). *8 keys to body–brain balance*. W.W. Norton & Company.

Scaer, R. C. (2014 [2001]). *The body bears the burden* (3rd ed.). Routledge.

Scaer, R. C. (2017). The neurophysiology of ritual and trauma: Cultural implications. In J. Gordon-Lennox (Ed.), *Emerging ritual in secular societies: A transdisciplinary conversation* (pp. 55–67). Jessica Kingsley Publishers.

Shears, D., Nadel, S., Gledhill, J., & Garralda, M. E. (2005). Short-term psychiatric adjustment of children and their parents following meningococcal disease. *Pediatric Critical Care Medicine, 6*, 39–43.

Shudy, M., de Almeida, M. L., Ly, S., Landon, C., Groft, S., Jenkins, T. L., & Nicholson, C. E. (2006). Impact of pediatric critical illness and injury on families: A systematic literature review. *Pediatrics, 118*(3), S203–S218.

Sonnenberg, A. (2017). Rituals in gastrointestinal endoscopy at the crossroads of shaman and science. *Endoscopy International Open, 5*(7), E627–E629.

Tomlinson, P. S., Harbaugh, B. L., Kotchevar, J., & Swanson, L. (1995). Caregiver mental health and family health outcomes following critical hospitalization of a child. *Issues in Mental Health Nursing, 16*(6), 533–545.

van der Kolk, B. A. (2014). *The body keeps the score: Brain, mind, and body in the healing of trauma*. Viking Books.

van Pelt, F. (2008). Peer support: Healthcare professionals supporting each other after adverse medical events. *Quality and Safety in Health Care, 17*, 249–252.

van Pelt, F. (2010). Medically induced trauma and compassion: Reflections from the sharp end of care. *Indian Journal of Anaesthesia, 54*(4), 283–285.

Yehuda, R., Engel, S. M., Brand, S. R., Seckl, J., Marcus, S. M., & Berkowitz, G. S. (2005). Transgenerational effects of posttraumatic stress disorder in babies of mothers exposed to the World Trade Center attacks during pregnancy. *The Journal of Clinical Endocrinology and Metabolism, 90*(7), 4115–4118.

Resources

Gordon-Lennox, J. (2017). *Crafting secular ritual: A practical guide.* Jessica Kingsley Publishers.
Gordon-Lennox, J. (2019). *Crafting meaningful funeral rituals: A practical guide.* Jessica Kingsley Publishers.
Karr-Morse, R. with Wiley, M. S. (2012). *Scared sick: The role of childhood trauma in adult disease.* Basic Books.
Levine, P. A., & Kline, M. (2007). *Trauma through a child's eyes: Awakening the ordinary miracle of healing.* North Atlantic Books.
MITSS became AHRQ Patient Safety Network (PSNet), https://psnet.ahrq.gov

Robin Karr-Morse is a family therapist in private practice in Portland Oregon—a role she has maintained for 35 years. She has designed and directed three state-wide programmes in Oregon, each of which focused on supporting vulnerable parents in the challenging task of raising healthy young children. Robin has published two books on the role of trauma in early brain development and its impacts on health and behavioural outcomes: *Ghosts from the Nursery: Tracing the Roots of Violence* (2014 [1997], Grove Press) and *Scared Sick: The Role of Early Trauma in Adult Disease* (2012, Basic Books). *E-mail:* robin@theparentinginstitute.com

Juan Carlos Garaizabal is a voice movement therapist, vocal coach, and vocal performer who also offers accompaniment through trauma releasing exercises (TRE) and bioenergetic analysis. Drama, singing, speech, voice, and movement have greatly enriched Juan Carlos' personal and professional life. He thrives on the creative edge between therapy and art. He has published 'The Presence of Voice' (2015), 'Tapices de Voz' (2017), 'Homeopathy and Voice' (2019), and 'No más boleros, gracias' (2020). *Website:* juancarlosgaraizabal.com *E-mail:* info@juancarlosgaraizabal.com

Jeltje Gordon-Lennox, MDiv, is a psychotherapist trained in body-based approaches and world religions. Her research and practice is influenced by her life experiences in conflict zones on several continents, in particular her work with the International Committee of the Red Cross. She has written five practical guides on secular ritualising, two in French and three in English. This collection continues the conversation on ritual and trauma started in *Emerging Ritual in Secular Societies: A Transdisciplinary Conversation* (2017, Jessica Kingsley Publishers). Jeltje lives with her husband and their two children in Switzerland. *Website:* gordon-lennox.ch *E-mail:* jeltje@gordon-lennox.ch

Dinka Community Case Study

Healing Post-Conflict Trauma Through Ritual

Alex Namu Kamwaria

Conflict in Sudan has existed for decades and persists in spite of several peace initiatives. The most affected regions include the Blue Nile, Nuba Mountains, and Abyei. In 2011, Sudan held a referendum that endorsed the independence of South Sudan. Peace in the South Sudan has remained fragile. Violence and civil war continues over the control of natural resources, such as the oil-rich fields in the Abyei area. It has been alleged that stakeholders in the petroleum industry have created conflicts among communities living around the oil fields. Most of these conflicts pit the Dinka against other communities such as the Nuer, Atuot, and Shilluk.

Due to the perennial conflicts, many Dinka victims have been displaced as refugees (see Fig. 1). Some have fled to peaceful countries, such as Kenya, Uganda, Tanzania, and the United States of America (USA). Wherever they sought refuge, the victims carry with them the traumatic experiences of the war. Violence, torture, murder, rape, persecution, and killings were some of the traumatising events (Lutheran World Federation (LWF), 2011). The large-scale brutality of war affects the victim's psychic, social, political, economic, and cultural dimensions. The Dinka victims of war have suffered significant trauma (see Fig. 2). The civil war may have ceased, but recovery from long exposure to the traumatising war experiences has been complex and challenging. Many victims are still preoccupied with the images of destruction and death, have lost sense of their basic trust or faith in society, and experience feelings of rage and revenge, helplessness, humiliation, and victimisation. Others have developed maladaptive social patterns, such as prostitution, drug abuse, domestic violence, or organised crime due to the experiences of war. As such, the psychological faculties that were damaged by war trauma need to be healed, and this healing usually occurs through the support of other people.

A. N. Kamwaria (✉)
Department of Social Sciences, Machakos University, Machakos, Kenya
e-mail: alexkamwaria@mksu.ac.ke

J. Gordon-Lennox (ed.), *Coping Rituals in Fearful Times*,
https://doi.org/10.1007/978-3-030-81534-9_7

113

Fig. 1 A young refugee shepherd herds cattle. A South Sudanese refugee herds her cattle and together with those of other families in her refugee camp. | © UNHCR/F. Noy CC BY-SA

Trauma healing implies the decrease of loneliness, mood improvement, sense of inner peace, a decrease in isolation, anger, and bitterness, and a decrease in feelings of animosity and hatred towards others.

Efforts to heal war trauma have subjected many Dinka victims to a new and yet unfamiliar western psychosocial trauma healing discourse, known as post-traumatic stress disorder (PTSD). Although the PTSD diagnosis has considerable value, focus on the Diagnostic and Statistical Manual (DSM) tends to ignore the fact that some Dinka traumatic experiences are profoundly culture-bound and, therefore, might not be adequately addressed within the Western treatment framework. The PTSD classification promotes the assumption that the victims moved from normal life to traumatic experiences, and then back again to normal life, hence the prefix 'post'. Given that many Dinka victims were born during the war period and are still experiencing war-related traumas, trauma is not 'post', but rather part of everyday life. The Dinka language does not have the vocabulary for an illness such as PTSD. Trauma entails experiences of life such as poverty, hunger, separation of families, and failure to perform necessary rituals. Thus, in order to effectively heal the post-conflict traumas among the Dinka, it is crucial to move beyond the PTSD-oriented therapeutic protocols and forge links with community-based healing systems.

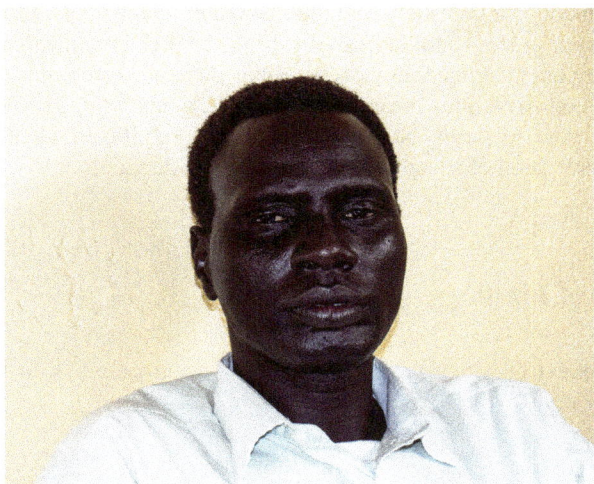

Fig. 2 Dinka man mourns loss of family to war. Alier Jacob was selected to be part of a programme to train 75 primary school teachers for seven months in a remote part of the country. His Nuer colleagues introduced him to their wives and children and invited him to their homes to break bread and discuss current affairs and history. They became like family. Thanks to them, he and his wife and children are safe. When the fighting started in Bor, he found out that people of the Nuer tribe had killed many of his (Dinka) family members. At first he became very angry; then he mourned for everyone who has suffered in this conflict. 'I have family and friends on both sides, Dinka and Nuer,' says Alier Jacob. 'When I think about what was done for me, I know that there are many good Nuer left'. | © Oxfam/Stella Madete CC BY

Study Methodology

The study employed cluster sampling that involved a three-tier sampling process. First, as researcher, I identified the ten States that constitute the Republic of South Sudan. Second, a selection of three States that had highest population of the Dinka victims of armed conflict was made. Third, in each of the three States, villages were identified. Convenience sampling was also used to select victims of armed conflicts. Then I worked with a team of research assistants who know the local language and culture of the Dinka community. In addition, purposive sampling was used to identify professional caregivers, indigenous healers, and respondents to be included in the focus group discussions. Finally, the snowballing method [1] was utilised to identify the ex-combatants.

---

[1] Snowball sampling is a research method whereby the researcher initially identifies respondents who are hidden or hard to find. The identified respondents are then requested to refer the researcher to other people in their category. The method is derived from the analogy of a rolling ball in snow that picks up more snow and becomes larger as it rolls downhill.

The participants for this study included Dinka ex-combatants, professional care-givers, refugees, elders, and indigenous healers (diviners, medicine men, spirit mediums). Although indigenous healers entail different categories, their roles are not mutually exclusive. This means that a healer can perform the roles of herbalist, diviner, and spirit medium. Notably, names and significant identifiers of study participants have been changed to protect their confidentiality.

Discussion of Findings

The social context in which the trauma actually occurred plays an integral role in understanding it.

Dinka Construction of Trauma and Healing

In essence, the Dinka cultural context creates the definition of trauma and appropri-ate healing. The study noted that the Dinka community has a vast majority of people who have lived most of their lives or were born during the period of the civil war. The armed conflict lasted for many years. Many of those who were taken as refugees or displaced returned to completely devastated villages, where houses were burnt, agricultural and grazing fields left derelict, and social amenities such as schools, hospitals, factories, roads, and railway lines destroyed. Many continued to live without the basic living conditions, such as food, clean water, and proper clothing. They had also lost their ties to family and community, the means to conduct social rituals, gainful employment, access to basic education, proper shelter, and other essentials of living.

Empirical studies on the physiological experiences of trauma across cultures show that virtually all therapies are effective when embedded within socio-cultural specificities of the victims (Kleinman, 1989; Green & Honwana, 1999). People differ markedly in their needs, reactions to experiences and situations, coping mechanisms, and resilience (Williamson & Robinson, 2006; Wessells & Monteiro, 2004). This implies that there is no single and universal way of treating post-conflict traumas. Indeed, it is the negotiability of a victim's traumas with his or her culture and worldview that would effectively heal the traumas. It is worth noting that although the majority of Dinka have converted to modern religious persuasions, their recourse to crises is by and large shaped by indigenous belief systems that are couched in traditional religion. Therefore, it is imperative to reframe the traumas experienced by Dinka victims of armed conflicts and devise healing mechanisms that would help them to repair their shattered lives and resume normal and productive lives.

Belief Systems of the Dinka

The Dinka consider a supreme being, Nhialac, to be the creator of the world. Though the Dinka vocabulary does not specify the gender category of Nhialac, more often than not, the creator is described in masculine terms. Nhialac has a will, emotions, and intelligence. This supreme being is all-powerful, all-knowing, all-beneficent, and just. After Nhialac come some divinities and spirits, who serve as functionaries in the theocratic government. Thus, divinities and spirits perform various duties in accordance with the will of Nhialac (Deng, 1980).

In the practical aspects of life, the Dinka are more concerned with their ancestor's spirits, clan spirits (*yieth*—singular: *yath*), and independent spirits (*jak*—singular: *jok*) than with a supreme being. The ancestors and *yieth* are partial and protective, while the *jak* are 'free' (subject to manipulation), malicious (may inflict injury or death), and stubborn (can enforce, reinforce, or sanction) (Deng, 1980).

Belief in a supreme being and other spiritual entities implies a certain type of conduct by the Dinka. At the centre of the Dinka code of conduct is morality. Any immoral act is considered a direct affront against Nhialac, the ultimate guardian of morality. The expectation is that Nhialac will punish evil acts. Immorality is redressed by a strict set of rituals that require the transgressor to undergo ritual purification and exculpation. Failure to do so leaves the transgressor vulnerable to socially contagious pollution known as *nueer* (Deng, 1980).

Post-Conflict Trauma Healing Among the Dinka

The Dinka notion of health is a state of complete wellbeing, based on a way of living, conduct, and behaviour in relation to the other members of the family and community (Deng, 2000). The notion of health imports respect to the dignity of the person and brings about the link of a person with a supreme being, divinities, ancestral spirits, community, and the environment. Health suggests harmony of body and spirit, while the lack of it implies disharmony between physical and cosmic forces. In this sense, trauma healing acquires a broader dimension that includes a harmonious relationship between a supreme being, people, and other extra-terrestrial entities.

Burton (1987) and Deng (1984) describe the Dinka as a community that places critical value on corporate or communal living and overall vitality of family, lineage, and community. Corporate social responsibility and mutual cooperation are emphasised in order to provide for the day-to-day basic needs of the community. Such an outlook ensures that all members of the community have their basic needs met, such as food and shelter, among others. Corporate living is anchored on the concept of *cieng*, which is underlined and ritualised in acts of friendship, mutual support, communion, and sharing. All actions, both in the sacred and secular realms, are screened through *cieng*. For this reason, *cieng* is the appropriate point of reference for framing traumas, and providing coping mechanisms and healing.

Cieng creates opportunities for the victims to mourn, support each other, share trauma stories and narratives, and find ways to address their immediate and long-term situation. Given that the PTSD approach is encumbered by many challenges, there is need to use *cieng* rituals to effectively deal with traumas in a holistic manner.

Ritual for Healing Ex-combatants

This study noted that indigenous rituals were used to initiate and mobilise the combatants. Part of this initiation and mobilisation was done in the bush; it entailed imparting special skills to the soldiers so that they could enter combat with supernatural help. Interviews with the ex-combatants revealed that spirit mediums assisted in disclosing the plans of attack of the enemies. Traditional medicine men ritually gave supernatural protection from bullets and the power to magically disappear from the enemy's sight. After the war, most of the combatants were disarmed and demobilised, but few could fit back in to village life. This was in part due to the fact that the ex-fighters still felt bound by the power of the rituals they had performed as combatants. [2]

As noted above, rituals for reintegration and demobilising ex-combatants fundamentally consist of cleansing the evil that afflicts the ex-combatants due to the horrible acts they committed during the war. Family members and the community must also participate in the rituals. It is during the course of the reintegration rituals that the ex-combatant is expunged and cleansed of the evil that clings to them due to involvement in immoral acts, such as killing, rape, torture, looting, and destruction of property. [3]

The head of the household starts the ritual by proceeding to the cattle byre and spending some time cajoling the ancestral spirits. He then selects an aggressive bull that is likely to bellow loudly while dying, and presents it to the ancestors. The bellowing connotes the approval by ancestors. The colour of the bull should be pure white, popularly known as *mabior*. [4]

When pressed about why a white bull was used, the elders said that the *mabior* has noteworthy roles. First, any man of substance, such as the head of a family, or community elder, is called *Mabior*. Indeed, every man has a bull's name that denotes his characteristic qualities. Second, the colour of the bull denotes purity. Third, the

---

[2] Unpublished interviews with Alier Mading Nyadeng, Gabriel Gai Kiir, and Macar Aweckoc Jiel conducted on 18 September 2015 in Lainya Village, Eastern Equatoria State, South Sudan (Kamwaria, 2015).

[3] Unpublished interviews with Manyuol Dong Mayadit, and Pajok Madingo conducted on 26 September 2015 in Torit Village, Eastern Equatoria State, South Sudan (Kamwaria, 2015).

[4] Unpublished interviews with Mayom Pajieng, Gilo Alemi, and Dan Bol Madut conducted on 26 September 2015 in Torit Village, Eastern Equatoria State, South Sudan (Kamwaria, 2015).

bull is part of family and community life, as a source of subsistence, wealth, and pride.[5] From that explanation, one can infer that the *mabior* is the best expiatory substitute for a Dinka man.

The study notes that before a bull is sacrificed, the ex-combatant rubs his body against it. Then the bull's throat is slit, and the ex-combatant jumps over it while it is still shaking and groaning before dying. There are several explanations for these ritual acts. First, the rubbing against and jumping over the bull symbolises transferring all evils, afflictions, or *nueer* to the sacrificial animal. Second, the death of the bull represents a locking away of the past. Third, the pouring of blood is meant to settle the angered spirits of the deceased so that they do not return to haunt the ex-combatant or his kinsmen.

Once dead, the bull is skinned and some of its parts separated and the ancestors given their part (through the burning of the meat to ashes), while the participants eat the rest. Some part of the blood is poured as libation to the ancestors, while the rest is sprinkled on the ritual participants. Thereafter, the ex-combatant is cleansed with water and smeared with red ashes of dung from the cattle that had eaten *apac*, one of the best pasture grasses in the Dinka compound.[6]

Commonly, the ritual consists of a series of symbolic enactments aimed at locking away the past. This is seen in the killing and jumping over the bull. The ritual helps establish an enduring and trusting relationship between the ex-combatant, his family, and the community. During the ritual the ex-combatant begs for forgiveness for the immoral acts or atrocities he committed during the war. He thus appeases the ancestors and divinities who will then protect him from any future afflictions from the angered spirits of those he killed.

Ritual After a Homicide

During the war, members of the community killed many of their own people. In Dinka tradition, the un-avenged spirit of the murdered person will haunt the murderer as well as the members of their family. Interviews with elders revealed that when a kinsman was murdered, justice was commuted by killing a kinsman of the killer with a social position equivalent to that of the deceased person. The goal of the reintegration rituals is known as 'anti-homicide'; that is, it is intended to prevent violent acts of revenge. The elders note that the rituals not only focus on the killer, but also on the families of both of the killer and of the person murdered.

In recounting the ritual, the elders explain how the two affected parties assemble and sit about twenty meters apart. The cattle that are to be paid in compensation to the family of the murdered person are placed between the two parties. A white bull

---

[5] Unpublished interviews with Mayom Pajieng, Gilo Alemi, and Dan Bol Madut conducted on 26 September 2015 in Torit Village, Eastern Equatoria State, South Sudan (Kamwaria, 2015).

[6] Unpublished interviews with Mayom Pajieng, Gilo Alemi, and Dan Bol Madut conducted on 26 September 2015 in Torit Village, Eastern Equatoria State, South Sudan (Kamwaria, 2015).

(*mabior*) must figure among the cattle offered in compensation. The ritual elder who officiates at the ceremony as the spear-master is perceived as a *jok* (independent spirit). The kinsmen of the killer seize the forelegs of the bull, and those of the person killed grasp the hind legs. Together they turn the bull on its back. Then one person from each side stabs the bull's chest with a spear. The bull is next cut into two halves. Its entrails are taken out and scattered over the two parties. Each party then goes away separately to divide their portion of meat. Parts of the bull are burned. After this, the *jok* goes to the spot where the bull was slaughtered, and places a spear among the remaining entrails. After making that gesture, he takes some of the remains and throws them over the two parties, who by this time have resumed their places.

The parties then advance in groups of six, three from each side, and they exchange the spears that were used to kill the bull. Next, they hold the spears between them, each person clasping it in both hands. Each spear-holder then bites on their spear. Spitting on each other follows this; they spit first on the person to their left, then on the one to their right, and finally downward upon their own chest. Then the *jok* sprinkles the ashes over the knees of the parties. This last gesture marks the end of the formal part of the ritual. After that the cattle that are to be paid out in compensation for those who were murdered are given out. No formal words are spoken. It is believed that after the conclusion of this ceremony, anyone who reopens the feud will surely suffer. [7]

A number of inferences can be made from this ritual. First, biting on the spear symbolises the willingness to abide by the settlement. Second, scattering of the bull's entrails, spitting, and dusting of the knees with ashes are forms of cursing the evil, purification, and blessing. Third, partaking of the meat signifies communion and unity (*cieng*). Fourth, killing the bull represents the characteristic notion of substitutionary atonement in this healing ritual. Although substitutionary atonement takes place during the public ritual, it is not complete until the aggressor truly asks for forgiveness and the aggrieved agrees to forgive. Fifth, healing goes far beyond the overt ritual to include mundane acts of judicial settlement through payment of cattle and a verbal commitment not to seek further retribution. Healing restores the relationship that was destroyed between the aggressor and the aggrieved, the community and a supreme being, divinities, ancestors, and other cosmic forces.

It is crucial to note that the aggrieved party is not obliged to forgive after the ritual. However, forgiveness is recommended since refusal to do so could enrage Nhialac, divinities, or ancestors, and invoke a curse. The aggrieved party should be ready to publicly acknowledge forgiveness and absolution for the ritual to be effective. The ritual prepares people to henceforth observe and uphold the moral values of the community.

Other interviews with victims of armed conflicts revealed that a pluralistic approach to healing after homicide was taken by some people. For example, after

---

[7] Unpublished interviews with Manyuol Dong Mayadit, and Pajok Madingo conducted on 26 September 2015 in Torit Village, Eastern Equatoria State, South Sudan (Kamwaria, 2015).

Fig. 3 Woman and children are particularly vulnerable in wartime. | © antheap CC BY-ND

the armed conflict some ex-combatants, who had been accused of killing many civilians, submitted themselves simultaneously to several healing strategies. Usually, they started with the Western psychosocial healer (for example, counsellors), and later moved on to indigenous healers (for example, diviners, spirit mediums, medicine-men) to complete their treatment.

Ritual After Rape

Baak[8] is an elderly Dinka woman who is respected by the villagers. During the interview, Baak described a ritual for cleansing the victims of rape (see Fig. 3). Since the Dinka equate cows with girls and women, a cow can be used for substitutionary atonement in cases of rape. Baak explained how the ritual elder starts by making invocations to Nhialac divinities and ancestral spirits. Then the rapist (if present), the victim, and close family members are assembled. After prayers, everyone goes to the river. The rapist and victim are submerged in the water and bathed. The sacrificial cow is also dunked into the same river for a few minutes. The animal is then killed, and its sexual organs cut in half.

Several deductions can be made from this ritual. First, the ritual provides an opportunity for the victims to be cleansed and purified. During the ritual, the perpetrator confesses the immoral act, swears never to repeat it, and asks for forgiveness. The ritual gives an opportunity for repentance and for assuaging shame and guilt for both the perpetrator and the victim. Second, the dunking of

---

[8] Unpublished interviews with Baak conducted by the author on 18 September 2015 in Bor town, Jonglei State, South Sudan (Kamwaria, 2015).

the rapist, victim, and sacrificial animal symbolises the cleansing of evil as it is passed onto the animal. Finally, the cutting of the sexual organs in half symbolically neutralises the immoral act of rape and returns the individual to their original status.

Implications for Trauma Healing

These rituals have a number of implications for the healing of trauma. First, the Dinka healing rituals present a holistic approach to health by combining both the social and physical dimensions of illness in order to treat the person as a whole. The Cartesian dichotomy on which PTSD is posited does not apply here where there is an overall integration of body, spirit, and mind. The social imbalance in a patient's life is generally reflected in the physical body; both social and physical dimensions are taken into consideration equally to restore the patient's health.

Second, the Dinka healing rituals revolve around the ideology of *cieng*. When one member of the community suffers, everyone else in the community suffers. Accordingly, when widespread calamity hits the community, such as civil war, the calamity is attributed to disrupted or broken relationships, which must be restored in order for the community to regain its harmony. Thus, the traumatised victim and his or her community become the starting point for diagnosis, prognosis, and treatment. It is noteworthy that the Western psychosocial trauma healing approach does not involve a concept like *cieng*. Rather, it physically isolates the patient from his or her family, companions, neighbours, community, and natural surroundings. When the patient is confined to a hospital ward, contact with family and community is limited and regulated. Moreover, family and community participation in the victim's healing is often limited to the payment of medical bills. *Cieng* healing rituals are holistic in the sense that they go beyond the individual patient to include the whole community in the healing process.

Conclusion

The Dinka notion of trauma is constructed in terms of social relationships rather than mental processes. For this reason, trauma healing in the context of internecine conflict requires factoring in the social dimension. The Dinka *cieng* rituals present a holistic approach to trauma healing by combining both the social and physical dimensions of trauma in order to treat the person as a whole. Trauma healing is achieved through a double strategy: social healing, which addresses with the supernatural causes of illness through rituals; and physical healing, which addresses relief of body pains through the use of herbal remedies. In addition, the Dinka healing system is interactive in that it involves not just the individual but the whole community. The rituals provide ample opportunity for the kith and kin in the community to participate in the diagnosis, prognosis, and treatment.

Some trauma victims integrate both the indigenous and Western healing approaches. The presence of professional counsellors and caregivers in the Dinka community attests to the fact that, on many levels, Western PTSD therapeutic approaches are utilised. Yet their acceptance is only superficial in that many patients routinely consult indigenous healers after undergoing counselling to complete their treatment, perhaps to determine 'who' caused the malaise and 'why' they were afflicted with illness.

References

Burton, J. W. (1987 [1954]). *A Nilotic world: The Atuot-speaking peoples of South Sudan.* Greenwood Press.

Deng, F. M. (1980). *The Dinka cosmology.* Ithaca Press.

Deng, F. M. (1984). *The Dinka of the Sudan: Case studies in cultural anthropology.* Holt, Rinehart & Winston.

Deng, F. M. (2000). Reaching out: A Dinka principle of conflict management. In W. Zartman (Ed.), *Traditional cures from modern conflicts* (pp. 95–126). Lynee Reinner.

Green, E. C., & Honwana, A. (1999). *Indigenous healing of war-affected children in Africa.* World Bank.

Kamwaria, A. N. (2015). *Integration of indigenous healing approaches into psychosocial model of trauma healing for Dinka victims of armed conflict in South Sudan.* Unpublished PhD thesis, Kenyatta University, Nairobi, Kenya.

Kleinman, A. (1989). *Patients and healers in the context of culture.* University of California Press.

Lutheran World Federation (LWF). (2011). *Focusing on the future of refugee.* Lutheran World Information.

Wessells, M., & Monteiro, C. (2004). Healing the wounds following protracted conflict in Angola: A community-based approach to assisting war-affected children. In U. P. Gielen, J. Fish, & J. G. Draguns (Eds.), *Handbook of culture, therapy, and healing* (pp. 321–341). Erlbaum.

Williamson, J., & Robinson, M. (2006). *Psychosocial interventions, or integrated programming for well-being?* World Health Organisation.

Alex N. Kamwaria, PhD in philosophy and religious studies, is dean of students and senior lecturer in the School of Humanities and Social Sciences at Machakos University, Kenya. Alex is the author of numerous articles and book chapters. He has also held several professional and administrative positions, including lead researcher at the Nairobi Peace Initiative—Africa (2005–2007) and academic dean in the School of Humanities and Social Sciences at Machakos University (2015–2017). Alex currently lives in Machakos, Kenya. *Website:* www.mksu.ac.ke *E-mail:* alexkamwaria@mksu.ac.ke

Memory Boxes

Ritualising Memory in Transitional Justice

Sophia Milosevic Bijleveld

> There are only very few things left from my three sons. These objects are very dear to me as they remind me of them. I am including them in the memory box to make my memories of my sons last. I want to keep them until the day I can finally ask the perpetrators why they killed my innocent sons. (Participant of the Memory Box Project)

Memory has long been a contested landscape. Especially in troubled times and places. Forty years of conflict in Afghanistan has left a trail of death, destruction, and pain. Despite their immense resilience, Afghan civilians are deeply traumatised; they have called for recognition of their suffering and for symbolic acts of reparation.[1] But, despite initial efforts, the transitional justice process has stalled today. Clearly, those who hold power are unwilling to document war crimes, prosecute these crimes, facilitate truth-telling, or even promote memorialisation that calls attention to the victims of the conflicts.

Over the years, war, ethnic and religious division, large-scale displacement, and lack of control of many parts of the territory by the Afghan central government have given rise to a multitude of narratives about the past and the conflicts. Each region, political fraction, or community interprets the past from their own perspective. The absence of a common historical memory has facilitated the repetition of catastrophic events. This is seen through how public spaces have been altered: museums or

---

[1] This was documented in the report by the Afghanistan Independent Human Rights Commission (AIHRC) entitled 'Call for Justice: A National Consultation on Past Human Rights Violations in Afghanistan' (2005).

S. Milosevic Bijleveld (✉)
Memorialist, Cultural Heritage Curator, Geneva, Switzerland
e-mail: smb@memorymatter.ch

monuments have been destroyed, and accounts of certain historical events have been manipulated or even intentionally omitted. Cases in point range from the destruction of Buddha statues in Bamiyan by the Taliban in late 2001, the omission of three decades of conflict in curriculum textbooks on Afghan history, and the statues and large posters that glorify perpetrators of horrific crimes.

Successive wars and foreign involvement in Afghanistan have led to profound changes in traditional social organisation, state structures, and interethnic relations. The party system in Afghanistan has not taken root. Continuous international intervention—which relies heavily on local militias—undermines the central government and leaves the political scene open to rule by ethnic group leaders. This context contributes to the fragmentation of collective memory; even a unifying discourse such as the Jihad in Afghanistan is fragmented by ethnicity. As Alonso (1988) emphasises, memory is the product of power struggles in which the dominant group uses its hegemony to interpret the past. This struggle to impose an official interpretation of the past is not fought on the frontline.

Battlegrounds of Memory

These power struggles are played out in Afghanistan on the battleground of memory. They are led by those with the means to build monuments and museums and to preserve or destroy archaeological sites. The elite display their claim to power over cultural heritage by erecting private museums, such as the Jihad Museum in Herat, which was instigated by warlord Ismael Khan. Other ethnic leaders assert their political power through the construction of monuments, both within their traditional stronghold and in the capital Kabul. Examples include tributes built for the late Hazara leader Mazari in Kabul and in Bamiyan, as well as monuments constructed for the late Northern Alliance leader Ahmad Shah Massoud in Kabul and in the Panjshir Valley. Celebrating heroes through the erection of such memorials reinforces the role of these actors in the memory landscape and further challenges and weakens the central government (Milosevic Bijleveld, 2011).

Collective memory, or the *mémoire collective*, a concept evoked by Halbwachs, is deeply linked to social interaction. The group or society to which an individual belongs is what allows him to form, retrieve, and give coherence and meaning to memories (Halbwachs, 1997). According to Halbwachs, memory is formed and constructed in this collective-specific framework. The individual, who retains only senseless images, needs the group and its framework to be able to recall his memories. By tying the concept of memory to the collectivity situates it in a time frame implying its limited duration. Halbwachs argues that memory is so connected to the group that if the group were to disappear the individual's memories would vanish. This highlights the importance of the contextualisation of memory.

According to Crane, lived experience and collective memory are interdependent (Crane, 1997). She emphasises the role of the individual in the process of collective memory. Indeed, even if she agrees with Nora that memory is preserved in the *lieux de mémoire*, or places of memory (Nora, 2001), Crane maintains that 'collective memory is ultimately located . . . in individuals' and that 'all narratives, all sites, [and] all texts remain objects until they are "read" or referred to by individuals thinking historically' (Crane, 1997, p. 1381). Zerubavel insists on looking beyond the individual to see collective patterns, and holds that this is best done through the sites of memory:

> Yet collective memory is more than just an aggregate of individuals; personal memories and such inevitably personal relief maps cannot possibly capture what an entire nation, for example, collectively considers historically eventful or uneventful. To observe the social marking of the past, we, therefore, need to examine the social timelines constructed by entire mnemonic communities. (Zerubavel, 2003, p. 28)

Memory is inherently linked to the present and to its systems of ideas. It relies on them to be consistent and understandable. This structures reality, and enables the group to shape and make sense of their memories. Memorialisation, or the process in which a memory of the past is transmitted, plays a central role in helping societies come to terms with the aftermath of atrocities and violent conflicts. Memorialisation calls for the ritualisation of memory through processes such as the renaming of streets, the use of religious or cleansing rituals, and the construction of memorials or museums. Beyond participating in a process of truth-telling and forms of symbolic reparation for victims, memorialisation involves reconciliation, encourages group cohesion, and involves the creation of a new national identity. Furthermore, as the victims participate—often spontaneously—in the ritualisation of memory, it brings a sense of healing to them, their community, and society as a whole.

The actors present on the battlefield of memory are not only the state or the elite. Since the fall of the Taliban, Afghan civil society has developed remarkably. Increasingly, the ritualisation of memory is being enacted by members of civil society, such as women, minority groups, or groups representing victims. In December 2009, the Afghanistan Independent Human Rights Commission (AIHRC), an independent body, opened a victims' museum in the outskirts of Fayzabad, the capital of Badakhshan Province. This museum, situated on the site of a mass grave, is the first of its kind in Afghanistan. The construction of the museum was initiated by the AIHRC with support from the local community. The museum has been visited by thousands of people. Today, it has the reputation of a shrine; rituals practised in intimate or special spaces are often superimposed on the visit to the memorial. Indeed, as visitors enter this secular museum, they remove their shoes, just as they would if they were entering a home, a shrine, or a religious building. Clearly, transitional justice is not unidimensional but consists of complex layers of approach. Moreover, as Celermajer (2013) underscores, for transitional justice to be effective and move beyond the legal framework, it must embrace the power of performative justice. [2]

---

[2] A performance-based relationship to justice involves tools such as the use of theatre in transitional justice.

Beyond the individual trauma and violence that Afghans have endured over these last decades of war, displacement, the ethnicisation of the conflict, and the radicalisation of traditional Islam have left profound marks on Afghan society. While memorialisation initiatives in Uganda or Cambodia relied upon traditional rituals for healing (Acirokop, 2010), Afghanistan seems to lack these kinds of healing rituals.

The Failings of Transitional Justice in Afghanistan

Under pressure from the international community to accept the Action Plan for Peace, the Afghan government reluctantly adopted a plan for justice and reconciliation in December 2005. This put transitional justice on the top of the agenda. The plan included recommendations from the AIHRC's report *A Call for Justice* about how Afghans wanted to deal with past abuse and human rights perpetrators. It outlined five key actions including symbolic measures, institutional reforms, truth-seeking, accountability, and reconciliation. It also clearly rejected the possibility of amnesty for crimes committed during the decades of war. It called upon then president Karzai to put in place a series of symbolic acts that would serve to recognise the plight of victims of war, such as the building of memorials or renaming public squares, designating days of remembrance, or other such acts which, while not costly, hold significant symbolic value.

Integral to the plan was an ambitious project known as the United Nations Conflict Mapping project, which involved reporting and documenting the abuses committed from 1978 to 2001 by all parties in the conflict. As the project evolved, it became a written report that resembled the work of a truth commission. The English version was completed in mid-2012; both the Dari and the Pashto versions were ready in late 2012. In December 2011, however, President Karzai took action against Nader Nadery, the commissioner overseeing the report, by dismissing him from his post. The report was briefly made available through the United Nations Assistance Mission in Afghanistan (UNAMA). Owing to concern about destabilising the country, the international community prevented this detailed and groundbreaking 1000-page Conflict Mapping report from ever being released.[3]

Following the publication by Human Rights Watch of a list of accused perpetrators, the Afghan parliament passed a law in 2007 (often called the Amnesty Law) granting blanket amnesty to 'all the political wings and hostile parties who had been in conflict before the formation of the interim administration'. Former President Karzai made a few amendments to the law that, while recognising the war victims' rights to seek justice, left the victim responsible for bringing charges against his aggressor before a court. In sum, this means that Afghan authorities may prosecute war criminals only on the basis of an accusation by a victim.

---

[3] Parts of the Conflict Mapping report were leaked online (see Resources).

This has effectively halted the transitional justice process put in place under pressure from the international community. As illustrated by other cases, such as Northern Ireland and Colombia, the engagement of civil society plays a powerful role in attaining and sustaining victim-centred peace. In Afghanistan, civil society has developed alternative approaches to this transitional justice process by implementing non-judicial mechanisms at the local level. These mechanisms, which are distinguishable from those implemented by the state or international actors, rely on non-judicial, innovative, and holistic approaches such as memorialisation.

Memory Boxes

A memory box resembles a small personal museum; it is built and curated by a family member in order to remember and commemorate the life of a person they have lost. The memory box consists of a small, portable wooden box in which the family member places significant objects and belongings from loved ones they have lost since 1978 to the conflicts and violence (see memory box displays in Figs. 1a, b, 2a, b and 3a, b).

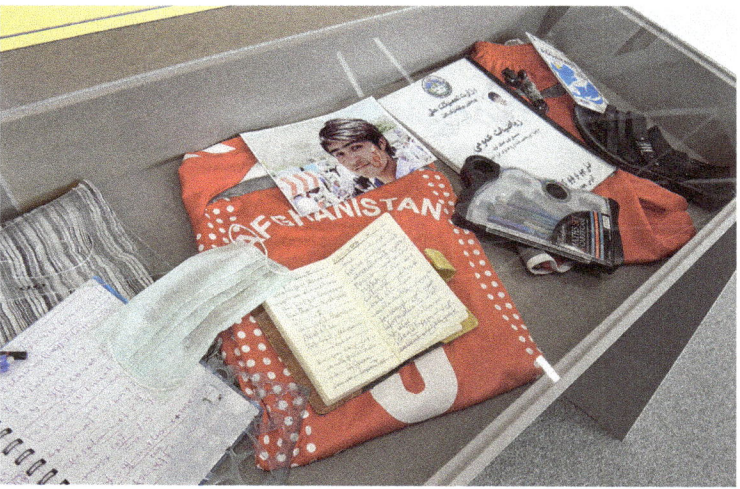

Fig. 1a Abdullah was 18 years old. 'It was his life-long dream to study engineering,' says Murtaza Darwishzadah who made this memory box for his brother Abdullah Frotan who was killed while attending a peaceful demonstration in Kabul on 23 July 2016. | © Philippe Palma

Fig. 1b Abdullah dreamed of becoming an engineer. Participants of the Memory Box Project create their memorial using their own words and objects under the guidance of a trained facilitator. | © Philippe Palma

Fig. 2a May my husband's memory live on forever. 'I have included these objects in my memory box to make my husband's memory live on forever,' says his wife Nargis. 'I want to share his story with the wider public. I also hope that in the future, the government will use these objects to build a museum for war victims. It will help the people of Afghanistan and the world to become aware of the pain and suffering our county and its innocents have endured, and to finally learn how to live in peace.' | © Philippe Palma

Fig. 2b The memory box keeps my husband's personal possessions safe. Nargis's husband Sakidad Hedayat, aged 33 years old, was tortured and killed while in detention in Kabul on 11 November 2000. | © Philippe Palma

Fig. 3a Everyone was sleeping when the explosion collapsed our home. Habib Wali made this memory box for his mother Rahilah (38 years old) and his sisters Tamana (16 years old) and Tabasum (11 years old) who died under the rubble of their home in Shah Shahid, Kabul at 1:15 am on 6 August 2015. | © Philippe Palma

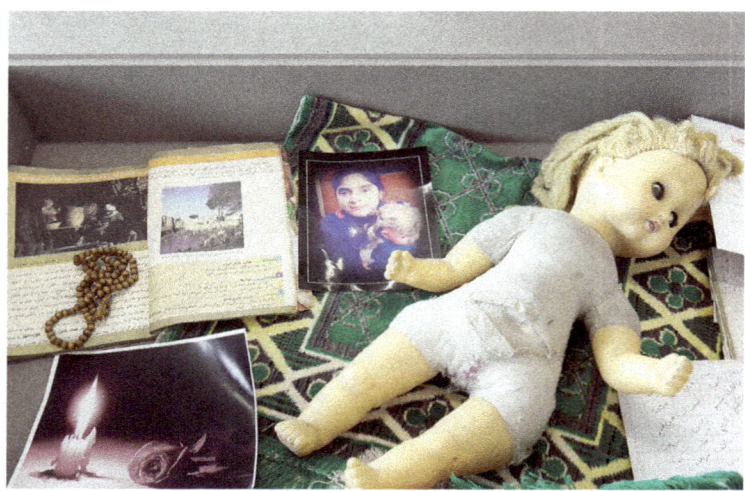

Fig. 3b My mother and two sisters died under the rubble of our home. For us, this was not the end of the story,' recounts Habib Wali. 'While we were out with our relatives holding a ceremony to mourn my mother and [two] sisters, thieves entered [the ruins of] our home and stole all that was left. We went to the police and to all other responsible authorities, but no one even tried to investigate or find the thieves. Let alone those responsible for the massacre.' | © Philippe Palma

The Memory Box Project, inspired by a European project from WWII, was adapted to the Afghan context by the local NGO [4] Afghanistan Human Rights and Democracy Organization (AHRDO) to address the decades of war and endemic violence as well as the lack of historical memory by bringing the victims and their voices to the forefront. In this innovative memorialisation project the victims ritualise their grief and loss through art, storytelling, and everyday objects to attain a symbolic sense of justice. The Project involves a psychosocial process that takes place during workshops that last from three to five days. It includes a full day of trust- and group-building exercises, as well as activities that gradually foster the participants' discovery of their creative capacities. A trained facilitator guides the participants through a variety of artistic games and activities to prepare them for storytelling. The participants, who represent different communities and regions, are asked to talk about their lives. All of the exercises are designed to foster dialogue and understanding, and to promote reconciliation amongst them.

Initial discussions, where the participants share their personal stories and experiences, are centred on war and peace in Afghanistan. Eventually, the focus shifts to making the individual memory boxes. Each participant chooses which moments of their life they wish to focus on and how to represent them in their personal boxes with specific images or objects. The boxes consist primarily of personal objects that belonged to their deceased family member, such as a copy of the Quran, a wedding

---

[4] NGO refers to non-governmental organisation.

certificate, jewellery, clothing, photos of deceased children, letters, and identity cards. During the creative process, participants may also produce new artistic pieces related to their current situation such as poems, paintings, or photographs of their home.

Another important element in this ritualisation process is the creation of a personal timeline. Each participant is invited to use coloured pens, crayons, and paint to sketch on paper the highlights of significant events in their life journey. They are also asked to reflect on the national flags of Afghanistan; as each new regime captured Kabul and imposed a new flag on the country, the flags represent important symbolic but also temporal markers that situate their life within the larger context of the war. Likewise, the imposition of a new flag is often associated with very painful personal losses. To reflect this, the participants are asked to paint the flag that is most significant for their own life story, and then draw another flag that depicts an ideal Afghanistan for them.

Once they have collected their personal objects and created the artistic artefacts (often as many as ten to fifteen objects per victim), the participants are invited to set up their own personal museum. They alone decide the most aesthetic and authentic way to assemble, curate, and prepare their memory box for display. Each of the participants then talks about the contents of their box. The psychosocial process concludes with a final walk through a symbolic Museum of Memory Boxes that closes with everyone sitting or standing in a circle to share their thoughts and feelings about what they experienced during the workshop.

The process of creating the boxes and sharing them with the public is painful and emotionally demanding for most participants. As the families ritualise their homage to the victim, these memory boxes become immensely precious to them. During the public display, the ritual becomes performative as family members guide and explain to visitors the significance of the objects they have presented in their memory boxes. Many of the participants speak of the healing benefits of being able to tell their stories and having others listen. Since the families often fear for the safety of their boxes, AHRDO has been entrusted with the safekeeping of the memory boxes.

Over the years, through its work with victim groups, AHRDO has collected over 200 memory boxes.[5] At the request of victims, a memory centre was opened to house the memory boxes and their narratives. The memory box display forms a timeline of the different stages of the conflict in the different regions; it also shows how the various ethnic groups and communities have been affected by the war. As a group, the individual memories and narratives in the boxes constitute a collective story. In this way, they contribute, collectively, in a truth-telling endeavour and to the healing process.

---

[5] The Afghanistan Centre for Memory and dialogued was inaugurated in February 2019, in Kabul, Afghanistan. It chose not to use the term museum in its title, as to distance itself from being a politicised space.

Conclusion

The Memory Box Project offers a unique perspective on the lives of victims that the visitor can explore through the reflective creative process of their personal stories. The creation of physical spaces of remembrance and memory enable the affected individuals and their communities to gain a better understanding of the conflict, come to terms with that violent past, and participate in reconciliation.

In the post-conflict context, memory plays an important role in building lasting and sustainable peace. Memorialisation is often narrowly referred to as symbolic reparation, but it goes beyond the brick and mortar memorials or monuments of physical spaces of memory to include grass-roots initiatives such as commemorations, the renaming of public places, apologies, or reburials. These efforts are increasingly central to transitional justice processes as ways of coming to terms or dealing with the past, but also as alternative approaches to truth-seeking and accountability.

The different memorialisation initiatives that have multiplied around the world contribute to a range of unconventional mechanisms for reconciliation, healing, and truth-telling. They also encourage individuals to partake in civic engagement in order to restore the social fabric of communities and societies that have broken down due to the violence. In order for memorialisation to have a profound lasting effect on society and deter violence, it must be an inclusive process that involves the victims; it must allow for various interpretations and reflections on the past. The experiences of the population at large are legitimate and have their place alongside the versions of history presented by the various political leaders.

In contexts of impunity, as is the case in Afghanistan today, people feel compelled to do something to ease their pain and loss. The ritualisation of memory offers unique ways for ordinary people to initiate and take an active role in their healing. The different aspects of the creative process of the Memory Box Project meet a vital need for people to heal their trauma and restore broken connections as they remember their loved ones. It aspires to help the victims of Afghanistan's conflicts assume an active role in the creation of public memories of the country's long history of violence and human rights abuses. Furthermore, the Project promotes the ideal of peace as a basic human right and recognises victims' memories as a central element of the peace-building process. Only by remembering the past can people hope to ensure that history does not repeat itself.

References

Acirokop, P. (2010). The potential and limits of *mato oput* as a tool for reconciliation and justice. In S. Parmar, M. Roseman, & S. Siegrist (Eds.), *Children and transitional justice: Truth-telling, accountability and reconciliation* (pp. 267–292). Harvard Law School in collaboration with UNICEF.

Afghanistan Independent Human Rights Commission (AIHRC). (2005, January 25). *A call for justice: A national consultation on past human rights violations in Afghanistan.* Accessed August 1, 2020, from www.refworld.org/docid/47fdfad50.html

Alonso, A. M. (1988). The effects of truth: Representations of the past and the imagining of community. *Journal of Historical Sociology, 1,* 33–57.

Celermajer, D. (2013). The role of ritual is shifting collective dispositions. In C. Brants, A. Hol, & D. Siegel (Eds.), *Transitional justice: Images and memories* (pp. 123–139). Ashgate.

Crane, S. (1997). Writing the individual back into collective memory. *The American Historical Review, 102*(5), 1372–1385.

Halbwachs, M. (1997). *La mémoire collective.* Albin Michel.

Milosevic Bijleveld, S. (2011). Forgetting, remembering: The jihad museum and the victims museum in Afghanistan. In *Museum of ideas: Commitment and conflict: A collection of essays* (pp. 310–339). MuseumsEtc Ltd..

Nora, P. (2001 [1984]). Entre mémoire et histoire. In P. Nora (Ed.), *Les lieux de mémoire* (pp. 23–43). Gallimard.

Zerubavel, E. (2003). *Time maps: Collective memory and the social shape of the past.* University of Chicago Press.

Resources

Afghanistan Human Rights and Democracy Organization (AHRDO) Website. Accessed on 1 August 2020 at http://ahrdo.org

Afghanistan Independent Human Rights Commission (AIHRC). Website. Accessed on 1 August 2020 at www.aihrc.org.af

Conflict Mapping Report. Accessed on 1 August 2020 at www.flagrancy.net/salvage/UNMappingReportAfghanistan.pdf

International Coalition of Sites of Conscience. Website of the only global network of historic sites, museums and memory initiatives that connect past struggles to today's movements for human rights. Accessed on 1 August 2020 at www.sitesofconscience.org

Memory Box Exhibition. (2015, December 10). Footage of a public event in Afghanistan capital Kabul, organized by AHRDO in commemoration of National Victims Day. Accessed on 1 August 2020 at www.youtube.com/watch?v=fMhShk4ROYo

Sophia Milosevic Bijleveld, PhD, is a memorialisation specialist who has worked most recently in Kabul at the new Afghanistan Center for Memory and Dialogue. Previously, she served as the global networks programme director at the International Coalition of Sites of Conscience. Sophia has extensive experience working with cultural heritage NGOs in Pakistan and Afghanistan, and as consultant curator in Rwanda and Kenya. Coming from a political science background, she focuses her research on the political use of cultural heritage, the politics of memory, and the political and social role of museums in post-conflict situations. She has published various articles on the role of museums and memory. *E-mail:* smb@memorymatter.ch

Networked Solidarity

Online Rituals of Mourning Following Public Death Events

Sasha A. Q. Scott

The public death event is an unfortunately familiar part of contemporary life. It involves those moments when the world's media are focused on the death of a distant victim that is seen as exceptional, morally significant, traumatic and worthy of public mourning and grief. The hyper-personalisation of networked media increases the scope and scale of these death events until a global audience experiences them collectively.

Our online responses to these deaths are explored in this chapter through the phenomenon that I term social media memorialising (SMM). SMM refers to the making and sharing of memorial videos, posting selfies of solidarity, the strategic use of hashtags, the remediation of symbolic imagery, and—perhaps most importantly—the collectives that form online around these digital expressions. These are episodes of repeated and simplified cultural communication, where prescriptive validity is drawn from the mutual experience of participants.

In other words, SMM is a form of spontaneous ritualising, performed because, in times of trauma, humans naturally seek emotional shelter, comfort, and escape in the community of ritual (Koster, 2003). In this chapter I examine SMM through the prism of death rituals by looking at how social network sites (SNS) can operate as ritual spaces in times of trauma. I argue that SMM serves an important role in a healing process that is in itself created by the mediation of these death events. A vernacular of grief is formulated through user generated content (UGC) and the architectural affordances of each platform. SMM invites participation; it is public, interactive, mediated, unscripted, and open-ended. These conditions produce powerful but ephemeral networked groups of otherwise disparate individuals that draw emotional strength and legitimacy from one another.

S. A. Q. Scott (✉)
European Broadcasting Union, Geneva, Switzerland
e-mail: scotts@ebu.ch

J. Gordon-Lennox (ed.), *Coping Rituals in Fearful Times*,
https://doi.org/10.1007/978-3-030-81534-9_9

137

The Media Death Event as Collective Trauma

Digital technology is impacting the way we grieve and mourn. It gives rise to new forms of expression that hybridise online and offline practices (Walter et al., 2012). The things people do online in the wake of traumatic events can feel strange, exaggerated, and even inappropriate to those of us whose formative years passed without smartphones and social media. Why do some people feel the need for public and highly emotional displays in response to a tragic news event? What do they get out of it?

At the heart of these questions lies our evolving relationship with networked technology. Digital technologies that capture and sharing content, the incessant pull of rolling news cycles, and the convergence of broadcast news with social media platforms bring these distant deaths of others close to us. The context is important here, as I am referring to the deaths of individuals that are considered to be exceptional, morally significant, and experienced as traumatic because they carry a sense of social injustice. This marks these deaths out from the online memorialising that follows the loss of personal friends or family on the one hand, and the deaths of celebrities on the other (see respectively Carroll & Landry, 2010; Haughey & Campbell, 2013).

Media and political actors have long ritualised the death event through days of official mourning, flags at half-mast, interruptions to scheduling, etc.—the majority of which are experienced via media. Thus the death event is also inextricably a media event (Dayan & Katz, 1992; Sumiala, 2014) and, by being performed and narrated through the symbolic and emotional representations of media, it is constructed as a trauma event. These events are *collective* because the death itself is experienced *in* public and *by* the public, and therefore our reaction is shared and communal (see Cottle, 2006; Couldry, 2003; Scott, 2017). We are no longer passive media audiences, but reformulated as mediated witnesses (Frosh & Pinchevski, 2009). As such, witnessing has two dimensions: that of seeing, and that of saying (Peters, 2001). This is the second piece of the jigsaw, for just as we witness these deaths via media, we turn to these same tools to respond. Internet platforms—and in particular social network sites—are increasingly defined by a type of sociality that blurs the distinctions between public and private, resulting in innovative, interactive, creative, and strategic responses to collective experiences such as public death events.

Death and Ritual Online

Death causes rupture. Unjust political death fractures individual and social identity, and can poison a communal sense of security. Our ritual responses to death contain vital questions of individual and collective identity, of intersubjectivity: these are explicit moments when we experience ourselves as unequivocally 'social' beings. Networked media has become embedded in our everyday lives at a startling rate, opening up new opportunities and forms of meaning-making. As such, online memorialising is in many ways a natural evolution, as technologies for publicising death have continually developed, from bell ringing to newspaper obituaries and now on to Facebook pages. Yet just because we can observe repetitive, stylised social actions in response to public deaths, this does not in itself determine the presence of ritual. After all, as Leach observed '[there is] the widest possible disagreement as to how the word ritual should be understood' (1968, p. 526).

Ritual can be highly elaborate, formal, and controlled, or it can be spontaneous and unstructured, and it is the latter that concerns me here. As such, and whilst resisting any potentially restrictive definition, I find it useful to think of ritual as constituted by three basic elements: some kind of collective gathering, a shared emotional experience, and a common focus on an object or action. The distinction between large social spectacles, such as football matches or concerts, and ritual is found in the intentions of the participants. For example, a football match is not ritual simply by definition, but it certainly holds the potential to become ritualised, particularly when it involves symbolically important opponents. The point is that ritualisation is the means by which one event is strategically separated out from the rest. This may not be easily grasped by the external observer, but is instinctively felt by insiders.

It has been suggested that this understanding of ritual is too dependent on the physical co-presence of human beings to be successful online; a vital 'ingredient' missing from the equation (Collins, 2004). However, there is a growing body of research that draws from anthropology, theology, and media studies that shows the reverse to be true: that media can host and cultivate powerful, meaningful and authentic rituals. Scheifinger's (2013) study of Hindu worship online, Connelly's (2013) exploration of Buddhist ritual in Second Life, Boyns and Loprieno's (2014) work on Interaction Rituals in parasocial relationships are just a few. I have previously written on the use of mobile telephone apps for religious ritual purposes (Scott, 2016). Colloquially, we often speak of an occasion having a 'sense of ritual', and this is very important: ritual is something we feel. We may well not share or agree on the specific details of a particular ritual, but we can share the sense of ritual that distinguishes a day as one of celebration, restraint, or reflection. Rather than trying to build a strict model of ritual, it is perhaps more productive to explore what ritual does in a given context.

A review of the literature illustrates how death rituals perform three major functions for a social group (see Durkheim, 1976 [1915]; Metcalf & Huntingdon, 1991; Sumiala, 2013; Turner, 1969; Van Gennep, 1960 [1909]). First, death rituals locate the dead within the physical and moral boundaries of the community, serving to confirm and strengthen social bonds amongst members. Second, death rituals aid transition: they are a bridge between dying as an event and death as a state of being. This transition concerns the dead, the bereaved, and the wider community of mourning, for all will be changed by the death event. Third, death rituals serve to heal, reconciling the bereaved with the loss they have borne and the future implications of that loss, and providing a framework with which to manage emotions and articulate grief. These categories are not mutually exclusive, and nor do they follow a neat linearity, but they serve an important purpose in understanding why death rituals to hold such social significance.

Demarcating Ritual Space Online

An important function of ritual is defining a space and the actions that take place within it as special, distinguished from the mundane workings of everyday life. Death rituals such as bathing and dressing the dead, sitting in mourning, the way in which a corpse is disposed of, the prayers and acts we perform as we mourn: these all operate within a strictly controlled and mutually acknowledged system of meaning that marks out the space in which they take place. This in turn informs certain norms of behaviour that are recognised and respected in some form or another. In traditional terminology we would call this 'sacred space'. Churches, mosques, temples, and graveyards all have a social inheritance through which a sense of 'sacrality' is predetermined, whether one is an adherent of that religious tradition or not. By association, the acts performed within these spaces are themselves, rather uncritically, perceived to be sacred. Yet this is an understanding of sacrality that is static, controlled, and highly restrictive. Indeed, in both academic and popular usage, the sacred is treated as little more than a synonym for the religious, and reflects a flawed assumption that this is some universal or cross-cultural phenomenon that we all agree on.

In the fluid, open, public spaces of SNS the symbolic systems of institutional ritual are absent, so the space is not demarcated as 'special' in any predetermined sense. This difference needs substituting or replacing in some form so that the participants of SMM and their interactions, and by definition the spaces in which those interactions take place, are framed in distinction from the usual noise online. This is not a question of replication (as in building a virtual simulacrum), but rather negotiating and constructing boundaries in the vernacular of the digital space. As Catherine Bell explains, ritual 'is designed and orchestrated to distinguish and privilege what is being done in comparison to other, usually more quotidian, activities' (1992, p. 74). That is to say, ritual is a strategic form of action that defines what, where, and who is special, which in turn informs and defines expectations of behaviour in context.

A social network site is never just one thing for one person: it is a homepage, an advertising space, a meeting place, a conversation, a performance, and more, all interwoven with one another. Yet these are also bound and restrictive spaces: terms and conditions are defined and policed by the company that owns it (or more often by its algorithms) and a page or group administrators; communicative acts are restricted by the structural architecture of the platform; community standards are negotiated and self-enforced by users. Thus, one of the most striking characteristics of SMM is the remediation of universal grief mnemonics to perform exactly this function: flowers, candles, angels, and stylised and iterative messages of condolence proliferate, quickly defining the space as one of mourning and grief. If the victim has a live SNS profile (in particular with privacy settings that allow anyone to post messages on their page), it can quickly become transformed from a personal performance of self into a collective performance of grief. For example, when extremists in London murdered the soldier Lee Rigby in 2013, his Facebook page became an important space for mourners across the globe to gather and share their thoughts, garnering hundreds of thousands of posts. Where such personal pages do not exist following a death event, a new Facebook 'Group' is often created, which then becomes a site for SMM. In Egypt, a single page called 'We are All Khaled Said' was set up in tribute to a young man brutally murdered by police, gaining hundreds of thousands of followers. In India, we saw a multitude of separate pages set up to commemorate Jyoti Singh Pandey, the high-profile victim of a brutal rape and murder on a Delhi bus in 2012.

YouTube is similar to Facebook but the architecture of the platform creates a different vernacular. Here, the video is the central focus of attention. Entering 'tribute Aylan [Alan] Kurdi'—the four year old Syrian refugee who died in the waters of the Aegean Sea in September 2015—into the YouTube search bar will return many thousands of results. The videos are hugely diverse yet carry remarkably similar visual and audio traits: flowers, hearts, candles, etc., generally soundtracked with sombre and melancholy pop music. Twitter is more abstracted; there is no single page or profile, and so ritual 'space' is an aggregate constructed between many users and connected through the inclusivity markers of a hashtag. Yet each shares the same aesthetic in their formulation, signalling to others the purpose of the space and thereby the norms of behaviour expected within it.

Formal arrangements of flowers and candles combined with an image of the dead are constitutive of a shrine across almost all cultures (Walter, 1996). As Drury (1994, p. 103) explains, candles and flowers 'are of particular importance as powerful symbols of remembrance, resurrection and immortality'. Flowers symbolise the fragility of the cycle of life; they communicate the need for nurture if they are to thrive, and they carry a particular aesthetic beauty in the emotional ugliness of death. Similarly, candles reflect an instability and temporality. These are simple yet powerful symbols, drawn upon and activated because they are accessible to such a wide audience: they are accessible and unambiguous representations of grief.

In digital spaces, flowers, candles, and portraits of the dead operate as grief mnemonics, becoming cognitive shortcuts to the larger event. They are often supplemented by other symbols, from the mundane (such as teddy bears and symbols of nature) to the religious. This is, in many ways, a direct recreation of the types of 'spontaneous shrines' that have become familiar to us at the physical sites of tragedy (see Santino, 2004). But the spaces of SMM are more than just shrines because they are co-constructed, interactive, and continually evolving. Online, the barriers to entry are far lower (you do not need to travel to the site, nor invest financially in an offering), existing in what Castells calls 'timeless time and the space of flows' (2001, p. 438). The sequential order of social practices is disrupted; they operate in spaces defined by users regardless of their physical location. As such, the online spaces invite a type of strategic participation that serves to co-produce the terms of engagement for SMM, in a manner that is far removed from the passive death rituals of organised religion that most of us are familiar with. The profile or content pages of SNS become 'virtual reliquaries' (Halverson et al., 2013), hosting traces of the dead with which we can interact in a timeframe and with levels of emotional immersion that are chosen and defined by us.

Symbolic Imagery and the Veneration of the Victim

Whilst grief mnemonics help define the ritual space, they do not hold all the answers. The public death event is formulated in and through digital imagery. We witness via technologies of capture and gaze, and that act of witnessing impacts our understanding of events. Thus, it is visual testimony of the victim that tends to take centre stage in SMM. The ability to appropriate and remediate the visual is a fundamental aspect of social media more widely, with found imagery acting as a shortcut in articulating an emotion (Ibrahim, 2012). As such, images of the victim are posted that generally conform to one of two types: (1) the death scene, communicating the injustice of events, or (2) the victim in life, in a venerative frame. As the two are often interspersed, the narratives of each reinforce one another.

Over time, these images tend to become increasingly abstracted from the original life stories. The victim that is endlessly remediated becomes 'flattened' into a reductive symbol, stripped of personal nuance. In other words, the victim moves from being a referential figure that denotes the dead person to a symbolic one that communicates something about the impact of the death event. As Assmann and Assmann put it, 'what had started as an image *of* ends up as an image *for*' (2010, p. 235, emphasis added). The victim becomes an icon: a cognitive shortcut that condenses meaning and emotion into an instantly recognisable form. Yet this also traps the victim in a liminal state of permanent 'just-dead', eternally replayed and revisited for our voyeuristic consumption.

SMM builds in layers of participation. In remediating symbolic imagery of the victim, participants become both more invested in the emotions of the moment, and mutually aware of each other. Boundaries of inclusion and exclusion are defined as the space becomes ritualised; the actions of the group become increasingly iterative and stylised. The vernacular of the digital—likes, up-votes, shares, retweets, comments, etc.—replaces embodied mechanisms in producing a shared emotional and cognitive experience, which then feeds back into the digital objects. Emotion isn't so much contained within these objects—they have no great or predefined importance—as it is a result of their treatment within the ritual.

The more these symbols circulate, the more affective they become. Collins (2004, p. 37) observes that when an object (i.e. image) comes to symbolise a collective emotion, a new 'sacred' object is born, but this feels compromised by its out-dated terminology. More convincingly, Kuntsman (2012, p. 6) suggests that we should think of the way these ephemeral objects become mediators and repositories of socialised emotional experiences in terms of the 'affective fabrics' of digital culture.

Locating the Self in a Communion of Grief

Ritual is a constant negotiation between the individual and the collective. When mourners physically gather in sacred space, signifiers—such as stylised dress (black, white, covered faces/heads, etc.), joining a procession, singing, etc.—all communicate our inclusion within the community of mourning. These encoded behaviours, rich as they are in tradition, create an awareness between participants. Each becomes more intensely focused on their common activity, more aware of each other's awareness, and thus co-producing a shared emotional experience. In many ways, a successful ritual can be understood as having produced moments of intersubjectivity, or what Durkheim called 'collective consciousness' (1976 [1915]). Yet these rules do not apply to the digital space, and so the self must be located within the collective body in new and innovative ways that gives rise to a new vernacular of grieving and mourning specific to each platform (Gibbs et al., 2015).

One such form is the 'selfie of solidarity': a stylised digital self-portrait combined with a shared political statement. A prime example is the ritualisation that followed the death of Kenji Goto, the Japanese journalist who was brutally murdered by ISIS in 2015. Across Twitter and Instagram individuals posted selfies of solidarity in the tens of thousands that carried the slogan #IAmKenji. By flipping the focus from the victim to the witness, the human body is visually inserted back into the ritual arena. In combination, the hashtag works as a label of identity and solidarity. We might easily draw comparisons between the use of the hashtag and mantras in traditional rituals, and with this understanding see how #IAmKenji functions to focus attention in a way that further diminishes the impact of the spatial remove between participants.

This is not to say that hashtag use *equals* ritual, any more than washing one's feet or taking wine always constitutes ritual. Instead, the hashtag represents a contextual combination of digital architecture and creative user behaviour, and serves to ritualise the space accordingly. If digital mediation does cause collective effervescence to be lessened, the increased scale of participation afforded by SNS seems to counter this. In using hashtags, individuals know their posts will potentially be viewed by hundreds of thousands of people because visibility is driven by the popularity of the hashtag and not dependent on their personal network.

Social media present us all with the means with which to insert ourselves into events and their aftermath in creative ways and with different degrees of immersion. Producing video content, up-voting, recommending or sharing, comments, and replies: these are all affordances that signify our presence both to the immediate group and to the distant, unknowable audience. This kind of communicative presencing takes many forms in comments threads, below videos, in updates and in tweets. The most common of these involves posting personal reactions and reflections that have echoes of offline memorials, but tend to express much more subjective and heightened emotions. They articulate a reverence for the death as an event that is conjoined with its witnessing audience and one step removed from the dead as an individual: it is about *us*, rather than *them*.

These highly stylised statements would be inappropriate if they appeared on our social platforms outside the context of SMM. When taken from its Latin root *precari*, meaning simply to make earnest petition, the statements can be seen as constituting prayer. The most common form is the simple and universal invocation for the deceased to 'rest in peace' (and the derivatives R.I.P, RIP, etc.). This combines with embodied identity acts as people add their location to the statement: RIP from Germany, Canada, Cyprus, the Philippines, etc. Here, Koster's observation that ritual is 'the symbolic demarcation of a territory in space and time' is particularly striking (2003, p. 214). These are all forms of *communicative presencing* that insert the individual into the narrative rendering of events. Scrolling down our screens, these iterative statements have a powerful visual impact: the repetition is inescapable, and the sanctity of the space is amplified.

Closely linked are the computational tools that serve as locative signals: hashtags, YouTube titles and descriptions, and Facebook pages are digital mechanisms that inform popularity, searchability, visibility, and promotion. Accordingly, they play a role in defining the parameters of inclusion for SMM. Within the pages of SNS, resources such as comment threads and their platform-specific vocabulary (likes, emoticons, up-votes, etc.) collapse the physical and conceptual space between users, creating a sense of emotional proximity, presence and, therefore, solidarity. Locative signalling both invites and helps construct the audience. Yet there is an inherent paradox within these rituals, because even though the collectives that are formed articulate powerful social bonds, once attention has shifted these are soon forgotten. These are not 'communities' as traditionally understood in sociology, inasmuch as members do not have any long-term commitment to one another. Instead, we should think of the participants of SMM as constituting a *communion of grief*: a temporary community, bound by mediated events and their ritualised reaction to them. Inclusion in this communion is defined by moral alignment, participation is open and actively encouraged, and validity and legitimacy are established through participation. This notion of a communion better reflects the intimate fellowship that is observed without implying a permanence of emotional bonds.

Conclusion: Networked Solidarity and Healing

SMM represents a coming together in emotional solace made possible by networked media, online platforms, and UGC. It is found in the expression of sympathy of Facebook, the hashtag of solidarity on Twitter, and the sharing of a tribute video on YouTube: small, potent, symbolic acts of anger, defiance, pain, and loss. They are stylised, iterative, and offer differing levels of emotional immersion. Individually, each act is diminutive, but in combination they can be immense, with millions of people contributing through the communicative affordances of social media: sharing, tweeting, posting, tagging, pinning, watching, liking, and streaming.

The ritual of SMM is defined by an absence or reconfiguring of institutional rules, officiating and authoritative figures, strictures of time and place, prescribed actions, and the voice of tradition. These are moments of creativity, spontaneity, even freedom of expression. The vernacular of online mourning is tailored to the material and user patterns of each platform, yet the iterative, routinised nature of SNS results in an aesthetic formalisation of ritual in both object and behaviour. Grief mnemonics create virtual shrines, and symbolic imagery links the victim with the political narrative of their death, creating mental shortcuts for our wider understanding of events.

The ritualisation of the digital space hosts a liminal communion of grief; a strategic, unstructured, and marginal mode of social relationship that contains the power to heal, console, and even absolve the empathetic guilt invoked in the mediated witness. Networked solidarity is the means through which trauma is diminished. Just as networked media brings the harrowing death of distant others close to us, so it provides the tools for us to push back emotionally, decreasing our affective proximity to—and thereby perceived responsibility for—the event in question.

The ability to enact and control social rituals is a powerful resource over which religious and state institutions have long held the monopoly. The spaces and scripts of ritual act as a system to recursively reproduce and structure control of social narratives of signification. In SMM, networked publics are seen to appropriate the process of public mourning, representing a reconfiguration of the power to assign and define who or what may be 'worthy' in today's world.

The media ecology of contemporary Western life results in an almost instantaneous remediation and veneration of the victim online. Our online spaces offer very different resources and conditions for framing the dead, not least because access to the means of production is open and interactive. In this way, the discourses of public death that SMM formulate do not just record or represent historical events, but they construct and constitute them. SMM does much more than sharpen a sense of solidarity: ritual creates narratives that are based on intensely felt moral definitions of the self and of the collective. As Durkheim first discussed, social rituals create a sense that you are part of something bigger than you could be on your own; that you are a dynamic part of history in-the-making; that your moral intuition is the right one and is shared by a collective body to which you truly belong. The digital circumvents ritual specialists and collapses barriers of time and space, accelerating and amplifying the scale of public mourning in unscripted, highly creative, and personalised ways.

References

Assmann, A., & Assmann, C. (2010). Neda – the career of a global icon. In A. Assmann & S. Conrad (Eds.), *Memory in a global age* (pp. 225–242). Palgrave Macmillan.

Bell, C. (1992). *Ritual theory, ritual practice*. Oxford University Press.

Boyns, D., & Loprieno, D. (2014). Feeling through presence: Towards a theory of interaction rituals and parasociality in online worlds. In T. Benski & E. Fisher (Eds.), *Internet and emotions* (2nd ed., pp. 33–47). Routledge.

Carroll, B., & Landry, K. (2010). Logging on and letting out: Using online social networks to grieve and to mourn. *Bulletin of Science, Technology and Society, 30*(5), 341–349.

Castells, M. (2001). *The rise of the network society: Economy, society and culture. Vol. 1 Information Age Series*. Wiley-Blackwell.

Collins, R. (2004). *Interaction ritual chains*. Princeton University Press.

Connelly, L. (2013). Virtual Buddhism: Buddhist ritual in second life. In H. Campbell (Ed.), *Digital religion: Understanding religious practice in new media worlds* (pp. 128–135). Routledge.

Cottle, S. (2006). Mediatized rituals: Beyond manufacturing consent. *Media, Culture and Society, 28*(3), 411–432.

Couldry, N. (2003). *Media rituals: A critical approach*. Routledge.

Dayan, D., & Katz, E. (1992). *Media events: The live broadcasting of history*. Harvard University Press.

Drury, S. (1994). Funeral plants and flowers in England: Some examples. *Folklore, 105*(1–2), 101–103.

Durkheim, É. (1976 [1915]). *The elementary forms of the religious life*. (J.W. Swain, Trans.). Allen and Unwin.

Frosh, P., & Pinchevski, A. (2009). *Media witnessing: Testimony in the age of mass communication*. Palgrave Macmillan.

Gibbs, M., Meese, J., Arnold, M., & Nansen, B. (2015). #Funeral and instagram: Death, social media, and platform vernacular. *Information, Communication and Society, 18*(3), 255–268.

Halverson, J. R., Ruston, S. W., & Trethewey, A. (2013). Mediated martyrs of the Arab spring: New media, civil religion, and narrative in Tunisia and Egypt. *Journal of Communication, 63*(2), 312–332.

Haughey, R., & Campbell, H. A. (2013). Modern-day martyrs: Fans' online reconstruction of celebrities as divine. In M. Gillespie, D. Herbert, & A. Greenhill (Eds.), *Social media and religious change* (pp. 103–120). De Gruyter.

Ibrahim, Y. (2012). The politics of watching: Visuality and the new media economy. *International Journal of E-Politics, 3*(1), 1–11.

Koster, J. (2003). Ritual performance and the politics of identity: On the functions and uses of ritual. *Journal of Historical Pragmatics, 4*(2), 211–248.

Kuntsman, A. (2012). Affective fabrics of digital cultures. In A. Kuntsman & A. Karatzogianni (Eds.), *Digital cultures and the politics of emotion: Feelings, affect and technological change* (pp. 1–15). Palgrave Macmillan.

Leach, E. (1968). Ritual. In D. Stills (Ed.), *The international encyclopedia of the social sciences*. Macmillan.

Metcalf, P., & Huntingdon, R. (1991). *Celebrations of death: The anthropology of mortuary ritual*. Cambridge University Press.

Peters, J. D. (2001). Witnessing. *Media Culture Society, 23*, 707–723.

Santino, J. (2004). Performative commemoratives, the personal, and the public: Spontaneous shrines, emergent ritual, and the field of folklore. *The Journal of American Folklore, 117* (466), 363–372.

Scheifinger, H. (2013). Hindu worship online and offline. In H. Campbell (Ed.), *Digital religion: Understanding religious practice in new media worlds* (pp. 121–127). Routledge.

Scott, S. A. Q. (2016). Algorithmic absolution: The case of Catholic confessional apps. *Online-Heidelberg Journal of Religions on the Internet, 11*, 254–275.

Scott, S. A. Q. (2017). Mediatized witnessing and the ethical imperative of capture. *International Journal of E-Politics, 8*(1), 1–13.

Sumiala, J. (2013). *Media and ritual: Death, community and everyday life*. Routledge.

Sumiala, J. (2014). Mediatization of public death. In K. Lundby (Ed.), *Mediatization of communication* (pp. 681–700). Walter de Gruyter.

Turner, V. W. (1969). *The ritual process: Structure and anti-structure*. Cornell University Press.
Van Gennep, A. (1960 [1909]). *The rites of passage*. (M. Vizedom & G. Coffee, Trans.). The University of Chicago Press.
Walter, T. (1996). Funeral flowers: A response to Drury. *Folklore, 107*(1–2), 106–107.
Walter, T., Hourizi, R., & Moncur, W. (2012). Does the internet change how we die and mourn? Overview and analysis. *OMEGA–Journal of Death and Dying, 64*(4), 275–302.

Sasha A. Q. Scott, PhD, is an interdisciplinary researcher, writer, and consultant in digital culture, media management, and digital transformation. He gained his doctorate from the University of London for his research on media events, ritual, and digital methods. Sasha is currently based in Geneva, where he is Project Lead for Digital Transformation at the European Broadcasting Union. Most of his free time is spent trying to keep pace with his three young children. *E-mail:* scotts@ebu.ch

Part III
Global Threat, Trauma, and Ritual

Challenging Global Dislocation Through Local Community and Ritual

Bruce K. Alexander and Matthieu Smyth

Addiction is one of the grimmest realities of the twenty-first century. Thousands of ragged 'junkies' and alcoholics in our cities are obsessed with getting the next fix of their substance of choice. They appear oblivious to the risks of contracting deadly illnesses, dying of an overdose, violence, suicide and all the other hazards of addicted life. They push themselves out of bed to faithfully perform the compulsive habits of addiction.

Yet street addicts using alcohol and illicit drugs represent only one small corner of the immense, doleful tapestry of human addiction. People of all social classes are severely addicted to habits of every sort, many of which do not involve drugs at all. Each one tries to satisfy their need with distinct, but repetitive, routines that structure their days and nights.

The Addict's Paradise

We find gambling addicts in the casinos and online, money and power addicts in the financial district, game and social media addicts on their computers, television addicts on their couches, bulimics at the junk food store, prescription drug addicts at the pharmacy, work addicts at their desk through the night, exercise freaks at the gym, 'shopaholics' chasing deals in real and virtual shops, love addicts in other people's beds, tobacco addicts on cancer wards, ideological zealots hatching plots,

B. K. Alexander (✉)
Department of Psychology, Simon Fraser University, Burnaby, BC, Canada
e-mail: bruce_alexander@sfu.ca

M. Smyth
Département des Sciences Religieuses, Université de Strasbourg, Strasbourg, France
e-mail: msmyth@unistra.fr

© The Author(s), under exclusive license to Springer Nature Switzerland AG 2022
J. Gordon-Lennox (ed.), *Coping Rituals in Fearful Times*,
https://doi.org/10.1007/978-3-030-81534-9_10

and on and on. Nearly any activity can become so addictive that it puts us and the people around us at risk. Those who strive ruthlessly for command positions in business corporations or governments, for example, too often quench their addiction to power and wealth by destroying the careers or even the lives of other human beings, cultural treasures, and the earth itself (Slater, 1980; Cramer, 2002; Polk, 2014).

Why are so many people dangerously addicted in the rich globalised world of the twenty-first century? Why has scientific medicine, with its dazzling successes, not brought addiction under control? What can be done to counter the ravages of addiction? I address these questions historically. A historical perspective can afford an unhurried examination of why addiction has always existed, why it appears to be spinning out of control just now, and what modern society can hope to do about it.

Defining Addiction

I use the word 'addiction' here in the way that it is traditionally defined in the English language: 'The state or condition of being [immoderately] dedicated or devoted to a thing, esp. an activity or occupation' (*Oxford English Dictionary*, 2010). This broad definition of addiction encompasses, at one extreme, all the dangerous drug addictions that society abhors and, at the other, the wondrous devotion of dedicated parents, lovers, and humanitarians. This definition of the term does not fit comfortably into what I refer to as today's 'Official View', which defines addiction far more narrowly as a relapsing brain disease caused by the exposure of genetically predisposed people to addictive drugs. I have reached the conclusion, on the basis of decades of face-to-face work with seriously addicted people and of immersion in the research of this field, that the medicalised, Official View of addiction as a brain disease is no better than the moralised view that preceded it in the public mind. The current Official View will eventually be seen as one of several dead ends in modern society's futile attempt to medicalise its way out of its immensely complex social and emotional problems (Alexander, 2008; Hart, 2013; Lewis, 2016; Vintiadis, 2017).

I argue that the fragmented societies, which we now all inhabit, are the most important cause of the flood of dangerous addictions that profoundly distort modern society.

Addiction: A Historical Perspective

Throughout pre-modern times, addiction in this broad sense was recognised—under various names—as a deadly pitfall for the human soul as well as the source of some of its wondrous events. In the worst of times, mass addiction was recognised as a factor in the downfall of cultures and empires (Alexander & Shelton, 2014). Many people—including myself—believe that we are facing exactly that kind of a ruinous situation today (McMurtry, 2013).

I have gradually become convinced that it was Plato himself who first formulated a comprehensive understanding of addiction (1987 [c.360]). In Book Eight of *The*

Republic (544a–575a) he contended that the root cause of widespread danger-ous addictions (which he called 'master passions') does not lie with the individuals who become addicted (moral weakness, disease, inheritance, or any other individual reason). Nor does it come from any special 'addictive' property of particular drugs or habits. Rather, Plato argued, the root cause of epidemic addiction lies in the structure of 'unjust' societies that are so distorted and imbalanced that almost everyone is drawn to one addiction or another (571a–573b).

Many historical studies corroborate that addiction is more prominent at particular periods of history and in certain kinds of society than in others. Indeed, the breakdown of functioning societies led to extensive dislocation, and to widespread addiction of many sorts, in the ancient world (Dodds, 1965), in aboriginal people who are being colonised and civilised, and in the modern world (Hughes, 1987; see review in Alexander, 2008).

Plato's insight on a particular kind of societal malfunction as the root cause of epidemic addiction and other psychological problems was almost completely for-gotten until it started reappearing in the writings of great social thinkers from the eighteenth through the twenty-first centuries. But why would such a brilliant social analysis of human problems, including addiction, be ignored for over 20 centuries, during which time no convincing understanding of the epidemic was found on the individual level? And how is it that this insight remains largely ignored today, when it sits in plain sight in a classic book that has been read in many languages by the most educated people? Further, why does the simplistic view that addiction is either an incomprehensible evil or an incurable brain disease still fill the media and the public consciousness?

Perhaps the real reason for not taking Plato's insight seriously is that to do so would force society to face terrifying realities about itself. When a deteriorating society reaches its final decline and is careening towards tyranny, it becomes very difficult for anyone to maintain a balance between their powerful, conflicting human needs. Unable to balance their needs, individuals adapt by focussing on a smaller number of intense needs, or even on a single need, in an effort to attain any kind of satisfaction at all. These overworked needs become addictions.

Plato explicitly invites us to face the nightmare possibility that in a society that is deteriorating in this manner, the most addicted people will become the political leaders. The ferocity of their addiction to power can make others turn to them in the vain hope of finding safety and a secure identity.

Fragmented Society and Mass Dislocation of Individuals

In the light of Plato's insight, our attention is not drawn to single individuals but to the societal causes of the rising tide of addiction over a five hundred years period that historians refer to as 'the modern era'.

Indeed, from the time of Christopher Columbus onward, all around the globe, colonising European nations crushed not only defenceless pre-modern societies and aboriginal tribes (Hobsbawm, 1989; Mann, 2011) but also many defenceless European rural cultures. Agricultural and industrial revolutions devastated stable peasant farms and commons in Europe. Refugees from this domestic fragmentation were cruelly stigmatised and economically exploited in slums or shipped abroad to populate the colonies.

European nations also fragmented their own elites. Manufacturers, and bankers competed relentlessly to maximise their individual wealth and glory, and some were ruined. Karl Polanyi (1944, p. 128) describes early modern England as being a place where 'the most obvious effect of the new institutional system was the destruction of the traditional character of settled populations and their transmutation into a new type of people, migratory, nomadic, lacking in self-respect and discipline—crude, callous beings of whom both labourer and capitalist were an example'.

Today too, the fragmentation of society escalates in wealthy and impoverished nations alike. The stringent economic rationality of the new major powers requires individuals to perform competitively and efficiently, unimpeded by ties to families, friends, traditional values, or norms of compassion. Advances in science and technology are no less to blame than the various forms of powerful ideologies such as Progress. These ideologies ingeniously provide justification for harnessing the entire planet to increase the wealth and power of the most 'civilised' nations.

The economic, political, and technological evolution that has bestowed enormous increases in industrial productivity and technical creativity on the human species makes it seemingly possible for the earth to support an emerging world civilisation of nearly eight billion people. However, this civilisation is in deep—possibly terminal—trouble, most obviously because of environmental destruction and obscene social inequalities, both of which are devastating side effects of fragmentation, as is mass dislocation.

Following Karl Polanyi (1944), I use the word 'dislocation' to describe the consequences of societal fragmentation on the psychology of individuals. The term refers to the experience of a void that can be described in many ways. On the social level it is the absence of sustaining connections between individuals and their family, local society, traditions, or even natural environment. In existential terms, it is the absence of vital feelings of belonging and purpose, which leads to disorders such as anxiety and depression. Countless literary works have described feelings of dread, absurdity, despair, loneliness, and void to lament the fragmented lives of the destitute, as well as those of the affluent and well educated. Evolutionary biologists speak of failure to satisfy the innate social needs of the human species in modern times. Dislocation is yet to be adequately described in the language of neuroscience, but research on the pro-social hormone oxytocin suggests a promising starting point (Merolla et al., 2013; Buisman-Pijlman et al., 2014; Alves et al., 2015).

Dislocation is more than just loneliness. In a fragmented society, even a frantic social life will not save a person from feeling the full force of dislocation (Lewis, 2016). Dislocation entails suffering that can afflict people in fragmented societies who remain within their community of origin as well as those who are driven continents away from their roots (Albrecht, 2012). Moving on to another place may offer advantages for individuals who feel stifled by their local communities, and provide opportunities for personal initiative or self-actualisation. However, prolonged radical dislocation exacts a high price that goes beyond today's obscene income inequality—which is but a toxic side-effect of the current economic system. Many wealthy people feel the full anguish of dislocation. No matter how rich you are, you cannot buy your way out of dislocation, although you may be able to create the appearance that you have.

Ultimately, dislocation generates unbearable misery in the form of anxiety, depression, disorientation, hopelessness, and resentful violence (Durkheim, 1951 [1897]; Mishra, 2017). Most of us cannot just 'tough it out'. This is why dislocation has been imposed as torture (in the form of solitary confinement or ostracism) from ancient times to the present.

The specific linkages between societal fragmentation and individual dislocation—along with the damage they cause—have been described at every stage of the cycle of life. Some of the brain mechanisms underlying this causal relationship have been worked out (Maté, 2008). Intrauterine consequences of stress endured by pregnant women in a fragmented society can make children feel socially ill-at-ease and, hence, dislocated, as they grow up. Lack of stable attachment in infancy or traumatic abuse due to fragmentation of families makes children feel insecure and less likely to achieve satisfactory integration in society later in life (Bowlby, 1969).

The opposite of dislocation appears harder to define. Following Erik Erikson, I use the term 'psychosocial integration'. I would suggest that it is the state of a person whose place in a well-functioning society enables them to feel that they belong, and yet still feel free. Although dislocation is the norm in a fragmented society, psychosocial integration is recognised as an ideal that some attain, at least occasionally.

Addiction as Adaptation to Dislocation

When nothing else seems to work, addiction can provide some much-needed compensation for a bleak, dislocated existence. Because addictive involvement only fills the excruciating void of dislocation partially and temporarily, addicted people must work their addictions for all they are worth, even if other parts of their life then fall to ruin (Alexander, 2008).

For decades I studied the surge of addictions amongst indigenous peoples in Western Canada, my own part of the world, which followed their forced resettlement on native 'reserves' by British colonisation. The devastating outbreak of harmful addictions that followed the fragmentation of indigenous cultures, and the consequent mass dislocation, is a tragedy that has been repeated with the same outcomes amongst aboriginal people on every continent that Europeans colonised (Mann, 2011).

To say that addiction serves a vital adaptive function is not to make light of it. Rather, it is to point out that, when faced with dire circumstances, addiction appears to be one of the only options for many people. That is why it is so common in our fragmented global society. Of course, addiction is not people's adaptive measure of choice. Nonetheless, in a fragmented world, addiction can provide severely dislocated people with some sense of belonging and purpose, at least in the short term. An addictive lifestyle can meet their desperate need for something to help them endure the crushing void of their existence. So, they cling to their addictions with the iron grip that they would apply to a piece of floating junk in a stormy sea. Quite often, they seize more than one addiction for safe measure. Without their addictions, most dislocated people would have little reason to live.

The adaptive function of severe addiction is often hidden. Many addicted people deny that they live in a state of dislocation because they feel ashamed of their inability to find a secure social life, a sense of who they are, or a reason to get up in the morning. In moments of insight, some can explain the function of their addiction with surprising candour; others may deny their dislocation because it feels unbearably like personal failure.

Mild and short-term addictions that involve behaviour that subsequently returns to normal may serve quite different adaptive functions. Burying oneself in work to finish an important project before a deadline, drinking one's way through a period of grief, or falling head-over-heels in love may have little or nothing to do with dislocation. You don't have to be dislocated to fall in love, but you do have to be dislocated to sacrifice your life addictively to dysfunctional love relationships (Peele & Brodsky, 1975).

Our ancestors adapted to their environment behaviourally and physiologically. We all survive by using the adaptive capacities that we inherited from them. The ability to be intensely dedicated to something is one of these. We are able to solve the most difficult problems by marshalling our energy and concentration in a sustained way, that is to say, by dedicating ourselves to finding the best solutions. Chronic dislocation creates a desperate need in human beings and other species that compete for survival as groups (Alexander & Shelton, 2014). Addiction is one of these adaptive capacities we use to endure bouts of dislocation. When addictions are short-lived and not too damaging we then move on to more socially integrated lives (Heyman, 2009). Some adaptive capacities that are obviously harmful from the outset are essential for survival because they protect a person from an even greater evil (limb amputation for medical reasons is one example).

Of course, dedication can go wrong when it is misplaced or when dislocated people have no other option but to continue their addiction beyond the limits of its effectiveness. Adaptation that facilitates survival at first can even become fatal if the dislocation is not relieved and the addictions become overwhelming, when adapting to a long-term stressor (Angeli et al., 2004). Adrenal stress responses known as the general adaptation syndrome, and the stress diseases that can result from these responses if they are worked to exhaustion, provide another example of this fact about adaptation (Selye, 1950).

If the partial utility of addictions to adapting individuals to dislocation explains their prevalence in a fragmented world, some also have adaptive functions for modern society itself. They help society to maintain high levels of production and consumption, which are required by our insatiable economy to keep the industrial wheels turning, the GDP growing, and the share prices rising. The addictions that favour economic growth and corporate profit—addictions to power, wealth, overconsuming, and overworking—are encouraged by modern culture.

There is still another reason why severe addiction is just as intrinsic to modernity, as competition or anxiety: the long-term consequences of severe addictions exacerbate fragmentation, thereby increasing dislocation and, ultimately, the prevalence of addiction. The vicious cycle takes another turn. Because of its long-term social consequences, severe addiction is not only a downstream response to social fragmentation but ultimately also an upstream cause of it.

With each new turn of the cycle, the flood of addiction rises to new heights and the costs to society increase. Consider the environmental and social destruction mandated by wealth and power addicts pursuing profits in the executive suites of their multinational corporations. Think of the wasteful consumption of millions of their addicted customers. Imagine the loss inflicted on society by talented people whose meaningless addictions get them stuck in a cycle of tenuous recovery and relapse. Not to mention the loss of the elders' stabilising accumulated wisdom to younger generations on account of the latter's addiction to television or prescription drugs.

To say that addiction is built into modernity is not to deny that it is affected by well-known risk factors such early life trauma, dysfunctional families, or predisposing genes. Rather, it is to say that the basic structure of modern society tilts the playing field in favour of addiction by ensuring that a huge number of people experience risk factors like these as the consequence of forces that are beyond their control, and by making it difficult for those who have acquired unsustainable addictive lifestyles to find satisfying alternatives.

Restoring Community: Friendship and Rituals

About three-quarters of the people who become addicted to a drug as young adults recover, usually without receiving any treatment. More than half of them recover by the time they turn 30 years old (Heyman, 2009). The heroin-addicted US soldiers who returned from the Vietnam War gave the most spectacular demonstration of this kind of 'natural recovery'; the majority of them recovered without any treatment. This should have dispelled the Official View for good.

Within the framework of the Official View, the high rate of natural recoveries from addition makes no more sense than the ineffectiveness of the War on Drugs (punishing drug users and traffickers) or the disappointing ineffectiveness of innumerable current forms of more compassionate treatment and prevention regimes. Billions of dollars have been spent trying to relieve individuals of the burden of addiction, and thus society of its consequences, to no avail.

The basis of natural recovery is no mystery. It occurs when people find psychosocial integration by establishing stronger relations with the community, or discover a strong sense of new meaning in life, thus reducing their sense of dislocation (Alexander, 2008). By contrast, both severe and minor addictions are much more common amongst people who are disconnected from the experience of their own inner richness and from a sense of having ties to family, community, traditional stories, and ritual that they see as their own.

The most successful solutions involve organising a less dislocated life on a local level. People who struggle with addiction restore community, in collaboration with other dislocated people, within self-help groups or local community groups. This is often referred to as the recovery movement; it provides community-oriented support, acceptance, and treatment to prevent and overcome addictions through small-scale social change. Rather than focus directly on addiction, it deals more broadly with dislocation, with or without the support of professional social workers or therapists to help people to find and retain a place in their community.

When addicted people overcome dislocation, they can leave their addictions behind. However, there is no easy way to help them to overcome dislocation, even in the context of today's recovery movement. First of all, because dislocation arises as it does from a deeply fragmented outer world, it can still break though to affect a less fragmented local society. Second, it is impossible to know in advance what will provide a basis for psychosocial integration in a particular group of people. Each person has a unique set of requirements for a fulfilling life. A local culture that effectively minimises dislocation cannot be established by fiat. It has to be carefully developed over a considerable period of time.

One of the best examples I know of the building of psychological integration is drawn from the community rituals of a young family with a troubled background whom I befriended a long time ago. Their psychological integration is due primarily to a successful endeavour to organise a stable group of friends through effective solidarity. It is also built around a series of secular rituals, such as Christmas breakfast for a dozen people, an annual party at the streamside to watch the salmon migration, and a 'Pie Day' at the peak of the blackberry season. Most remarkably, every six weeks for almost a decade, they have hosted a 'disc night' with about two dozen friends and their children. People bring food and drinks, as for a normal party. However, at about nine o'clock, all gather around the fireplace to hear a short speech introducing the evening. Sometimes it is followed by another that specifically presents the music to be played. People sit on the floor and remain quiet when the music begins. I could say the fireplace has analogous functions to that of the altar standing at the centre of a liturgy. When the music is over, people begin talking again, but softly. Eventually they drift back to their homes. These rituals have been fundamental to this family's efforts to cope with chronic dislocation (Alexander, 2008).

It seems difficult to restore those much-needed feelings of attachment, belonging, identity, and sense of self without meaningful, intensely cohesive rituals. Even more than the collective narratives that are always associated with ritual ceremonies in a traditional society, such collective ritual experiences allow people to feel part of a wider human collective experience of being in and of the world. Furthermore, what is lived through these integrative rituals is not a theoretical knowledge, but an existential experience; and an experience into which the individual story ultimately may fit.

Addiction is not a disease but, as in the case of any suffering person, self-soothing strategies can go terribly wrong and their social horizon can shrink drastically. In the struggle for recovery, people may be so taken up with their own traumas, failures, and need for solace, that they become addicted to 'recovery'. Overcoming addiction implies recovery from personal dislocation and, ultimately, from its source: social fragmentation.

We must not overlook the importance of community rituals in the prevention and recovery process. Rituals are part of every human social experience. Cohesive rituals, along with friendship, take an essential part in social bonding, and offer a powerful experience of belonging, identity, meaning, and purpose that shape every human community. Such cohesive rituals are essential to social bonding.

Cohesive rituals do not challenge addiction directly, but go beyond the healing practices that focus principally on soothing the pain of dislocation without confronting its source. Many healing practices, such as meditation, tend to encourage intense focus on oneself in the present moment—an approach quite ill-suited to addicted people who already tend to be intensely focussed on themselves. Living a full life depends on being aware of past history and the challenges of the future as informed by memory. The rituals and friendships vital to authentic social bonding transform the fragmented environment inhabited by the addicted person who can then recover a life where dislocation is kept in check.

As Marc Lewis points out, 'recovery' may not be quite the right term. Overcoming addiction does not involve returning to a previous state of being. Rather, it is about building a new outlook and a supportive local culture from the wreckage of the past (Lewis, 2016). We know how to put into place the elements that support individual and social wellbeing. But, once we have done all we can, we must then sit back and wait for the results of recovery groups to appear—and they often do.

Conclusion: Confronting the Colossus

At the end of the day, I fear that organising communities with the purpose of achieving a flourishing social life still focuses too narrowly on individual addicted people. Restoring people's place in a functioning community is much more difficult than it seems. Of course, small-scale social change groups do help many people. But, despite their goodwill, these groups alone cannot overcome the flood of addiction that engulfs modern civilisation.

Nor can this flood be stopped simply by restoring ancestral tribal traditions. These cultures in their pure forms have for the most part been wiped out or rendered obsolete by a highly technologised planet where most of its nearly eight billion inhabitants dwell in cities of mixed ethnicity. While 'retribalising' society is an essential part of the solution to this vicious cycle, it is nonetheless a hugely complex undertaking.

Yet once we realise that the causes of severe addiction are built into modernity itself, it becomes clear that our best hope lies in greater emphasis on large-scale social change. The clearest statement of the need for macro-level change that I know is contained in a legend first told to me by a native grandmother who was also a drug counsellor for her people. Drug counsellors from her tribe in Northern Canada sit by the side of a raging mountain river and watch, she says. When they see people being swept away in the white water they jump in to rescue them. They know how to find a path through the rapids to the drowning swimmers because their elders have told them where the rocks are hidden. Using all their strength, they eventually reach the drowning person and drag them through to the shore. With the counsellor's last ounce of strength, they heave the sinking swimmer up on the bank. When it is too late, the effort is wasted. The rescued swimmer slips right off the riverbank and is lost again in the foam. But sometimes the swimmer stands up and walks from the riverbank into the forest to re-join their people and return to the land.

When that happens, the storyteller told me, the drug counsellors feel like great warriors and they swell with pride. They would like to feel that theirs is a huge contribution... except that some 'son-of-a-bitch' upstream is throwing more and more people into the water all the time! Eventually the counsellors realise that, for all their heroic efforts, they are not winning but losing.

I believe that the heroic rescue work must continue. This involves taking into account our human need for cohesive rituals and their related stories, for both are meaningful to us. However, an even more vital task is getting rid of the son-of-a-bitch that is still lurking about. In other words, finding a way to jettison the whole vicious cycle that underpins addiction. At some point, when we face the fact that there truly is a son-of-a-bitch upstream, we will also have to confront the colossus of the modern age itself.

References

Albrecht, G. (2012, August 7). The age of solastalgia. *The Conversation.* Accessed August 2, 2020, from http://theconversation.com/the-age-of-solastalgia-8337

Alexander, B. K. (2008). *The globalization of addiction: A study in poverty of the spirit.* Oxford University Press.

Alexander, B. K., & Shelton, C. P. (2014). *A history of psychology in western civilization.* Cambridge University Press.

Alves, E., Fielder, A., Ghabriel, N., Sawyer, M., & Buisman-Pijlman, F. T. A. (2015). Early social environment affects the endogenous oxytocin system: A review and future directions. *Frontiers in Endocrinology, 6*(32), 1–6.

Angeli, A., Minetto, M., Dovio, A., & Paccotti, P. (2004). The overtraining syndrome in athletes: A stress related disorder. *Journal of Endocrinological Investigations, 27*, 603–612.

Bowlby, J. (1969). *Attachment. Volume I of attachment and loss.* Basic Books.

Buisman-Pijlman, F. T. A., Sumracki, N. M., Gordon, J. J., Hull, P. R., Carter, C. S., & Tops, M. (2014). Individual differences underlying susceptibility to addiction: Role for the endogenous oxytocin system. *Pharmacology, Biochemistry and Behaviour, 119*, 22–38.

Cramer, J. J. (2002). *Confessions of a street addict.* Simon & Schuster.

Dodds, E. R. (1965). *Pagan and Christian in an age of anxiety: Some aspects of religious experience from Marcus Aurelius to Constantine.* Cambridge University Press.

Durkheim, É. (1951 [1897]). *Suicide: A study in sociology*. (J. A. Spaulding & G. Simpson, Trans.). Free Press.

Hart, C. (2013). *High price: A neuroscientist's voyage of self-discovery that challenges everything you know about drugs and society*. Harper Collins.

Heyman, G. M. (2009). *Addiction: A disorder of choice*. Harvard University Press.

Hobsbawm, E. J. (1989). *The age of empire: 1875–1914*. Vintage.

Hughes, R. (1987). *The fatal shore: The epic of Australia's founding*. Knopf.

Lewis, M. (2016). *The biology of desire: Why addiction is not a disease*. Public Affairs.

Mann, C. C. (2011). *1493: Uncovering the new world Columbus created*. Vintage.

Maté, G. (2008). *In the realm of hungry ghosts: Close encounters with addiction*. Knopf Canada.

McMurtry, J. (2013, October 4). Corporate child abuse: The unseen global epidemic. *Global Research*. Accessed August 2, 2020, from www.globalresearch.ca/corporate-child-abuse-the-unseen-global-epidemic/5352919

Merolla, J. L., Burnett, G., Pyle, K. V., Ahmadi, S., & Zak, P. J. (2013). Oxytocin and the biological basis for interpersonal and political trust. *Political Behavior, 35*, 753–776.

Mishra, P. (2017). *Age of anger: A history of the present*. Farrar, Straus & Giroux.

Oxford English Dictionary. (2010). Addiction. In *OED Online* (3rd ed.). Accessed August 29, 2021, from https://www.oed.com/view/Entry/2179

Peele, S., & Brodsky, A. (1975). *Love and addiction*. Taplinger.

Plato. (1987 [c.360]). *The Republic* (2nd ed. revised). (D. Lee, Trans.). Harmondsworth: Penguin Classics.

Polanyi, K. (1944). *The great transformation: The political and economic origins of our times*. Beacon.

Polk, S. (2014, January 18). For the love of money. *The New York Times Sunday Review*. Accessed August 2, 2020, from www.nytimes.com/2014/01/19/opinion/sunday/for-the-love-of-money.html?ref=opinion

Selye, H. (1950, June 17). Stress and the general adaptation syndrome. *The British Medical Journal, 1*, 1383–1392. Accessed August 2, 2020, from www.ncbi.nlm.nih.gov/pmc/articles/PMC2038162/pdf/brmedj03603-0003.pdf

Slater, P. (1980). *Wealth addiction*. Dutton.

Vintiadis, E. (2017, November 8). The current medical consensus of addiction may very well be wrong. *Scientific American Blog Network*. Accessed August 2, 2020, from https://blogs.scientificamerican.com/observations/is-addiction-a-disease

Bruce K. Alexander, PhD, is a psychologist and professor emeritus of the Department of Psychology at Simon Fraser University in British Columbia, Canada, where he has worked since 1970. His primary area of research has been the psychology of addiction. His early 'rat park' experiments helped demonstrate that simple exposure to narcotic drugs does not cause addiction. More recently, he focuses on the causes of global addiction, not only to drugs but also to a great variety of habits and pursuits. He is married with four children and three grandchildren. See his book *The Globalization of Addiction: A Study in Poverty of the Spirit* (2008, Oxford University Press). *Website:* brucekalexander.com *E-mail:* alexande@sfu.ca

Matthieu Smyth, PhD, is a ritual anthropologist at the University of Strasbourg and trained as a somatic experiencing practitioner (SEP). He is the father of three children, an avid alpinist, and the author of *La Liturgie oubliée* (2003, Le Cerf) and *Ante Altaria* (2007, Le Cerf). Matthieu lives in Besançon, France. *Website:* rituelprimal.com *E-mail:* msmyth@unistra.fr

Ritual in an Age of Terror

From Taliban to Trump

Lisa Schirch

Humans turn to ritual and symbol in the best and worst of times. Rituals use symbols and symbolic actions to convey powerful messages of both hate and division, and love and connection. People use rituals to escalate and to respond to violence.

From Nazis to white supremacists to the Taliban, groups intent on restraining human rights and freedoms use rituals and symbols to foster hate and division. Nazis used massive military parades and swastikas to consolidate public support for the Holocaust and World War II. The Taliban orchestrate symbolic public stonings of women accused of adultery (Graham-Harrison, 2013). White supremacists burn crosses and carry torches in hate parades in the US (Morlin, 2017). Terror groups also exploit significant symbols. On 11 September 2001, Al Qaeda attacked two symbolic buildings, the World Trade Center and the Pentagon, which represented the heart of Western capitalism and military power. In France, ISIS attacked the cartoonists at Charlie Hebdo on 7 January 2015 as a symbol of Western lack of respect for Islam.

Ordinary people too can use rituals to respond to violence. In some cities, people create shrines of flowers and candles where terrorism takes place. Cellist Karim Wasfi performs at the site of a car bombing in Baghdad to infuse music and beauty at the site of violent trauma (Malone, 2015). Peace activists in Afghanistan walk across the country to plea for an end to fighting (Mashal, 2018). Community members line up to give blood in New York City hours after the 9/11 attacks (Starr, 2002). An African American professor leads her students through a Christian ritual of lament and transformation the day after the 2016 US presidential election (Cleveland, 2017). A group of witches cast a 'binding spell' on Trump to make it more difficult

L. Schirch (✉)
Toda Peace Institute, Tokyo, Japan

Alliance for Peacebuilding, Washington, DC, USA

George Mason University, Arlington, VA, USA
e-mail: lisa@toda.org

for him to harm people through his policies (Gault, 2017). In each of these cases, symbols and rituals are responses to violent extremism and the terror left in its wake.

This chapter explores ritual in an age of terror and violent extremism. It defines and compares the ritualistic aspects of terrorism and violent extremism. Then it documents and describes two ritual responses to the perceived violent extremism evident in the 2016 election of Donald Trump.

What Is Violent Extremism and Terrorism?

In 2015, nearly 30,000 people died in terrorist attacks (Global Terrorism Index, 2019). Despite news media bias against Muslims, terrorism does not correlate with any specific religion. Christians, Jews, Buddhists, Muslims, and atheists all carry out acts of terror. Terrorism is a violent *action*; a tactic aimed at inducing public fear and intimidation to achieve a political goal (Schirch, 2016).

The term 'violent extremism' refers to the set of *beliefs* that underlie the logic of *acts* of terrorism. The characteristics of violent extremism include a broad narrative that suggests some groups of people 'pollute' society and that these groups must be removed or killed in order to create the violent extremists' ideal vision of a 'pure' society. Violent extremism often includes the belief that violence against civilians is necessary to 'purify society'; the belief that people who are different racially, ethnically, religiously, or ideologically are inferior; the belief that women are property to control; the belief in authoritarian decision-making, and opposition to democratic practices of participatory decision-making (Schirch, 2018). Violent extremist beliefs encourage, condone, justify, or support terrorism.

Ritual Aspects of Terrorism and Violent Extremism

While some groups use rituals to carry out acts of terror, other groups harness ritual to resist violent extremist beliefs and respond to terrorism. Ritual has three specific characteristics. First, ritual occurs in a unique social space, set apart from everyday life. Second, ritual communicates through symbols, embodied actions, senses, and emotions rather than relying primarily on words or rational thought. Third, ritual confirms and transforms people's worldviews, identities, and relationships with others. Rituals are symbolic physical actions that form and transform worldviews (Schirch, 2005).

Violent extremist movements from Trump supporters to the Taliban use ritual as a way of enforcing division and fuelling hate. Acts of terror have ritual-like qualities. Terrorism is a unique event using shocking brutal acts of violence that take place not on a battlefield, but in the midst of civilian towns and cities. Acts of terror often target prayer groups, cartoonists, bankers, mall shoppers, and kindergarten children, rather than soldiers. Through dramatic symbolic acts that invoke strong emotion, terrorism sends the message that no one is safe. Terror groups preach that ritual acts of violence help to purify society. Terrorism aims to change and disrupt the status quo and bring about a new world order through ritual violence.

Terror groups use symbols and rituals to amplify their narratives that seek to justify cleansing society of people they deem impure. Some Buddhist monks in Myanmar use public prayer rituals as an opportunity to denounce Rohingya Muslims as a 'foreign enemy', setting the stage for genocidal violence against them. Some Jewish Israeli settlers plant symbolic flags on hilltops to claim land, and call for violence against Christian and Muslim Palestinians. Torch-bearing American young white men in Charlottesville, Virginia in 2017 marched around Confederate statues and shouted threats denouncing Jews, recent immigrants, and African Americans. Dylan Roof, a young white extremist, attended a prayer service at an African American church and opened fire on the church members. The massacre was an act of terror, but also a symbolic act. The white extremist chose the sacred space of a church to perform a symbolic killing. At his trial, Roof claimed the act of terror against this church prayer group was a symbolic act aiming to communicate hatred of African Americans, and with the intent of stirring up a 'race war' to create an all-white society (Sanchez & O'Shea, 2016).

US President Donald Trump is an example of a leader who propagates violent extremist beliefs. His speeches and tweets assert that immigrants and people protesting systemic racism are threats to his vision of a 'pure' society. His red caps announce that he can 'Make America Great Again', which many interpret as an affirmation of white supremacy. He blames COVID-19 on China, he bans or limits Muslim tourists and immigrants from the United States, and his immigration policies aim to punish migrants fleeing war and poverty. The Trump administration uses ritual performances to communicate their violent extremist worldview. Building a wall along a portion of the United States—Mexico border provided a symbol and performance of anti-immigrant views. During the protests against police violence in June 2020, militarised police forces teargassed peaceful protestors from a park near the White House to clear the space to walk to a photoshoot of Trump brandishing a Bible in front of the parish house of a church. Religious leaders widely condemned Trump's manipulation of religious symbols in his effort to undermine the movement to defend Black lives.

Americans have responded to Trump's election and espousal of violent extremist narratives and policies in a wide variety of ways, including the careful use of ritual and symbol to assert the humanity and dignity of all people.

Ritual Responses to the Election of Trump

While some groups use ritual as a tool for terror, other groups harness the power of ritual to resist violent extremist beliefs and to respond to terrorism. In the US, the election of Trump created a sense of fear, depression, and crisis. Some responded to the 2016 US election with ritual.

Reports from mental health therapists across the country highlight the growing anxiety among Americans following the 2016 presidential election (Clarridge, 2017; Miller, 2016). So many therapists were struggling with how to respond that some started a group called 'Citizen Therapists for Democracy'. The group's goal was to encourage therapy practice that addresses the public stress experienced both by the therapists and their clients by helping them keep their emotional balance, enact their values, and avoid toxic polarisation during the 'Trump era' and beyond. These therapists sought to resist anti-democratic ideologies and practices through positive, democratic engagement as an overarching value (Doherty, 2020). Within this context, ritual serves as a tool to address the anxiety stemming from Trump's violent extremist worldview threatening immigrants, people of colour, Muslims, and other groups targeted by his policies.

Christena Cleveland, an African American Christian theologian, describes a ritual pathway out of the endless cycle of violence. She reminds those who are white that the life of a woman of colour has never been safe or secure. So the election of Trump does not rupture reality in the same way it did for white women. '[F]or people of colour, this [election of Trump] is yet another trauma on top of a lifetime of traumas' (Cleveland, 2017). African Americans have a full repertoire of rituals including music, liturgy, and prayers to respond to trauma; they know how to use this tradition to respond to Trump's violent extremist agenda.

In a class held the day after the 2016 US election, Cleveland taught her students at Duke Divinity School a series of ritual practices she had designed. She describes a transformative ritual with stages for 'lament', 'truth', 'humanization', 'connection', and 'formation' (Cleveland, 2017). In the *lament* stage Cleveland asked students to name the injustices involved in Trump's election. These included:

> The victims of sexual abuse who now have a President who brags about sexual assault… Muslims who now live in even greater fear… people… incarcerated under an even more unjust US Justice Department… the silence of white liberals who should have spoken out about Trump in their conversations with their loved ones who supported… [and] the 81% of white evangelicals who turned their backs on Jesus and voted for Trump. (Cleveland, 2017)

Describing the process with her students, Cleveland says: 'The list went on and on. There were many tears, many wails of anguish, and many Amens' (2017). Next, the professor asked her students to reflect on the *truth* of who God is. These included God as 'compassionate, the God of justice, a mother bear who fiercely protects her cubs, a lover who knows our worst pain, the one who came and comes to a world full of strife, oppression, genocide and injustice, etc.' Cleveland explains that, 'Speaking truth about God strengthens us by expanding our hearts and enabling us to receive the good gifts of God, even in the midst of distress' (2017).

Humanisation is the third stage. Cleveland explains, 'Praying for the oppressor is an incredibly effective way to humanize the oppressor. If we fall into the trap of dehumanizing our oppressor, we inadvertently dehumanize ourselves—for our humanity is interconnected' (2017). Her students prayed for Trump, prayed for his staff by name, and prayed for those who supported and elected Trump.

Connection to the 'source' came next. Students journaled, prayed, practised a walking meditation, or sat in silence to connect with their understanding of the Creator or Divine Spirit. Each of these activities provided an opportunity to symbolically connect through an embodied action.

During the final class portion, Cleveland led students through a ritual process of *formation*.

> [W]e discussed what kind of leaders we want to be in this age of division, hopelessness, and fear. And we each created an individualized ethos—an aspiration to guide us away from our shadow self's less self-actualized responses to pain and disillusionment, and toward a higher standard of leadership... we each identified three tangible and accessible practices to help support us as we seek to lead from our ethos. Since I tend to be intense, overcommitted/ hyperactive and solemn when I'm afraid, I chose practices that would help me to play, connect with nature, and be still. My students committed to practices that would empower them to instigate difficult conversations with loved ones, participate in local direct-action protests, preach more boldly, see and bless the humanity of 'the other', and more. (Cleveland, 2017)

The ritual process Cleveland offers her students allows them to resist, protect against, and transform despair and stress. It occurs in a unique social space, set apart from everyday life. Participants communicate in this ritual process through symbols, embodied actions, senses, and emotions rather than relying primarily on words or rational thought. Finally, the ritual process confirms the humanity of all involved, and invites a transformation of participant's worldviews, identities, and relationships with others. Cleveland points a way out of a culture of thinking that violence or despair is the only solution to violence.

A Spell to Bind Trump: Magic Resistance

'Resistance witches' are another group of people using ritual to respond to Trump's election (see Fig. 1). They include around 13,000 people practicing Wicca and a variety of other magical or neopagan traditions. Some are practising Christians or former Christians, and they note the similarity of traditional Christian prayers. Every month since Trump's January 2017 inauguration, this group performs a spell that resembles religious ritual and activist performance. The spell is known as #MagicResistance.

Mainstream news media repeatedly criticise Trump's lack of facts and outright lies. Some have called Trump a 'post-fact' president. Resistance witches see magical rituals as an appropriate response. One of the originators of the spell, Michael M. Hughes, noted: 'Desperate times demand magical measures.... Trump's presidency [is] surreal and abnormal, therefore there [is] a need to counter him and resist his administration beyond the normal channels like public protests, petitions, emails, and calls to representatives' (Burton, 2017). For Hughes, magic should be part of the resistance to Trump's extremist agenda (Hughes, 2018). In response, Magic Resistance offers a modern-day 'binding spell'.

Fig. 1 Witches' mass ritual. | © Karina Ruiz Diaz @Kitty_Lemiew

Binding spells have a long history in the world of magic. They do not seek to hurt or harm Trump. Rather, they attempt to limit Trump's power so that he will not harm others with his policies. Activists carry out the binding rituals during the waning moon in ceremonial spaces in their own homes, in front of the Trump Tower in New York City, and in the forest in parts of America. These resistance witches express a relief to 'channel feelings of powerlessness about the current administration, while reviving a sense of community and ritual many report missing from their daily experience' (Hughes, 2018).

Defiant's Matthew Gault (2017) published the bind spell, later shared on *Medium*, asking others to perform the prayer/spell at midnight on every waning crescent moon until Trump leaves office. People can modify the prayer/spell according to their own spiritual practice or experience. The spell includes the following symbols arranged in

Fig. 2 Tower tarot card.
| © Matteo Villani

THE TOWER

a circle: a white candle to represent the element of fire, a small bowl of water to represent water, a small bowl of salt to represent Earth, a feather to represent the element of air, a piece of pyrite (fool's gold), and a shorter orange candle, a carrot, a Cheeto, or a photo to represent Trump. A Tower tarot card (see Fig. 2), which represents chaos, liberation, and troublemaking, is optional.

The spell or prayer includes the following *actions* (in italics) and words (in regular print).

Light white candle
Hear me, oh spirits
Of Water, Earth, Fire, and Air
Heavenly hosts

Demons of the infernal realms
And spirits of the ancestors
Light inscribed orange candle stub
I call upon you
To bind
Donald J. Trump
So that his malignant works may fail utterly
That he may do no harm
To any human soul
Nor any tree
Animal
Rock
Stream
Or Sea
Bind him so that he shall not break our polity
Usurp our liberty
Or fill our minds with hate, confusion, fear, or despair
And bind, too,
All those who enable his wickedness
And those whose mouths speak his poisonous lies
I beseech thee, spirits, bind all of them
As with chains of iron
Bind their malicious tongues
Strike down their towers of vanity
Invert Tower tarot card
I beseech thee in my name
[Say your full name]
In the name of all who walk
Crawl, swim, or fly
Of all the trees, the forests,
Streams, deserts,
Rivers and seas
In the name of Justice
And Liberty
And Love
And Equality
And Peace
Bind them in chains
Bind their tongues
Bind their works
Bind their wickedness
Light the small photo of Trump from the flame of the orange candle stub and hold carefully
above the ashtray
Speak the following loudly and with increasing passion as the photo burns to ashes
So mote it be!
So mote it be!
So mote it be!
Blow out orange candle, visualising Trump blowing apart into dust or ash
Pinch or snuff out the white candle, ending the ritual

Using the power of words that Trump himself has used, the spell can include the words 'you're fired' directed at Trump at the end.

This binding ritual shares common characteristics with Cleveland's ritual. Participants create a unique ritual space set apart from everyday life. The binding ritual communicates through symbols, embodied actions, senses, and emotions rather than relying primarily on words or rational thought. Then, the binding ritual confirms and transforms people's worldviews, identities, and relationships with others.

Whether or not the ritual actually works to prevent Trump's policies from taking place is, of course, not something that can be proven one way or the other. In the months after Trump's inauguration, the President did seem to be 'bound'. And certainly the laws in place, the checks and balances of the US constitution, the immense resistance by civil society, and these binding rituals may all have contributed to Trump's lack of progress in creating a more divided and hateful world.

In the online article *Art + Activism + Magic: Answers to Questions About the Mass Ritual to Bind Donald Trump*, Michael M. Hughes (2017) states: 'Magic has many definitions, and it works on many levels—inner, outer, cultural, practical, and artistic. Magic is art and art is magic—the two are inextricable.' Hughes also responds to some of the criticisms of the spell. When asked why he and other witches are not putting their energy into demonstrations, calling Congress, and sending money to the American Civil Liberties Union, Hughes responds: 'What makes you think we aren't doing that too?' In response to a question whether 'all this spell stuff is just a way to make yourselves feel good', Hughes retorts:

> What's wrong with doing things to feel better? As I previously stated, the spell also functions as a self-exorcism—a way to banish the nasty, oppressive Trumpiness out of one's mind and soul. And according to hundreds of people who have shared their experiences, it did just that. And that's magic, don't you think? (Hughes, 2017)

Some witches and pagans oppose the ritual because of their belief in the 'Threefold Law' or the 'Law of Return'. They say that anything negative you put out comes back to harm you threefold. Hughes responds that a spell to infuse Trump with well-wishes just does not seem appropriate (2017).

The Future of Ritual Response to Terror

Ritual resistance to Trump gained momentum in the spring and summer of 2020. Following the murder of an unarmed African American man, George Floyd, Black Lives Matter protesters gathered at Confederate statues across the US, denouncing these symbols of white supremacy. The mayor of Washington DC had the four-lane street beside the White House painted with the words Black Lives Matter. This was a performative ritual reclaiming the space after Trump had ordered forceful removal of protestors so that he could symbolically hold a Bible to suggest religious justification for his 'law and order' aggression toward protestors. White and Black protestors in hundreds of towns knelt in remembrance of Floyd in a symbolic nine minutes of silence (see Fig. 3). The ritual of kneeling was a symbolic act of apology, remembrance, and commitment to end white supremacy across the US. The ritual

Fig. 3 Taking the knee for George Floyd, Harrisonburg, Virginia. Ritual kneeling is a sign of respect and honour for the life of George Floyd, but it is also done in protest of racial injustice and police brutality. | © Lisa Schirch

communicated through a symbolic body posture and silent witness that Americans were committed to addressing racial injustice. These ritualised protests have already brought policy proposals and changes to policing, though they are the beginning and not the end of transformation in the US.

Ritualists like Cleveland bring liturgy and ritual to address despair at the election of a president who propagates violent extremism. Resistance witches draw on centuries of magic in a binding ritual that resists a political agenda that terrorises people and the planet. If terrorism chooses symbols and uses rituals to bring destruction, those who care about human rights, democracy, and freedom also can harness the power of ritual to affirm these values.

In her article 'The Politics of Public Mourning after Terror Attacks in France', Johanna Sumiala asks questions about how people respond to terror. In response to the 2015 attack on the cartoonists at Charlie Hebdo in Paris, a meme and Twitter hashtag #jesuischarlie, or 'we are all Charlie', was part of the response from the global digital community. But some contested this public mourning. While acknowledging the tragedy, some felt Charlie Hebdo was sexist and racist and did not want to express open affiliation with their message. Some cautioned that there were more Muslim victims of terrorism than white ones. A year after the 2015 attack on the Bataclan theatre in Paris, there were no sirens, bells, or fists in the air. Rather, there was silence. A silent memorial. No words. No speeches. Just the names of the dead, a marble plaque, the French flag, candles, and flowers. News media and social media shared photos, creating a 'digital community of mourners' (Sumiala, 2016).

Humankind is never quite sure how to respond to terrorism. Putting out flowers, reading names, or setting up a memorial are simple rituals. Some find giving blood an effective, meaningful ritual. But blood banks ended up throwing out excess blood after 9/11 because so many people wanted to participate in this ritual and relatively little blood was needed (Starr, 2002).

Artists like Karim Wasfi, the renowned conductor of the Iraqi National Symphony Orchestra and a cello player, reclaims bomb sites in Iraq by offering his music to channel and express emotion: 'I want this to be a global initiative against insanity and the impact of instability' (Malone, 2015). The horror of terrorism and authoritarianism is not easy to capture with words. We can move out of terror by going beyond the limits of rationality and verbal responses in our frontal cortex to explore how to better use our senses and emotions in ritual pathways.

References

Burton, T. I. (2017, October 30). Each month, thousands of witches cast a spell against Donald Trump. *Vox*. Accessed August 1, 2020, from www.vox.com/2017/6/20/15830312/magicresistance-restance-witches-magic-spell-to-bind-donald-trump-mememagic

Clarridge, C. (2017, March 24). Mental-health therapists see uptick in patients struggling with post-election anxiety. *Seattle Times*. Accessed August 1, 2020, from www.seattletimes.com/life/wellness/mental-health-therapists-see-uptick-in-patients-struggling-with-postelection-anxiety

Cleveland, C. (2017). Wellness in the age of Trump and terror. Accessed April 15, 2019, from www.christenacleveland.com/blog/2017/2/wellness-in-the-age-of-trump-and-terror

Doherty, W. J. (2020, January 20). My Journey as a citizen therapist. *Journal of Humanistic Psychology*, 477–487.

Gault, M. (2017, February 23). Use this spell to bind Trump and his cronies. Open-source magick targets the Donald. *Defiant*. Accessed August 1, 2020, from https://medium.com/defiant/use-this-spell-to-bind-trump-and-his-cronies-a5b6298f5c69

Global Terrorism Index. (2019). *Institute for economics and peace: Study of terrorism and response to terrorism*. College Park, MD: University of Maryland. Accessed August 1, 2020, from http://visionofhumanity.org/app/uploads/2019/11/GTI-2019web.pdf

Graham-Harrison, E. (2013, November 25). Public stoning consideration is latest setback for Afghan women's rights. *The Guardian*. Accessed August 1, 2020, from www.theguardian.com/world/2013/nov/25/public-stoning-womens-rights-afghanistan-government-adultery

Hughes, M. M. (2017, March 3). Art + activism + magic: Answers to questions about the mass ritual to bind Donald Trump. *Medium*. Accessed August 1, 2020, from https://medium.com/@michaelmhughes/art-activism-magic-answers-to-questions-about-the-mass-ritual-to-bind-donald-trump-11e52f94c23e

Hughes, M. M. (2018). *Magic for the resistance: Rituals and spells for change*. Llewellyn Publications.

Malone, B. (2015, May 28). Interview: Why I played the cello at a Baghdad bombsite. *Al Jazeera*. Accessed August 1, 2020, from www.aljazeera.com/news/2015/04/interview-played-cello-baghdad-bombsite-150429191916834.html

Mashal, M. (2018, June 15). A grass-roots Afghan peace movement grows, step by step. *New York Times*.

Miller, S. G. (2016, December 13). 1 in 6 Americans takes a psychiatric drug. *Scientific American*. Accessed August 1, 2020, from www.scientificamerican.com/article/1-in-6-americans-takes-a-psychiatric-drug

Morlin, B. (2017, April 10). Cross-burnings still a 'tool of fear' used by racists. *Southern Poverty Law Center (SPLC)*. Accessed August 1, 2020, from www.splcenter.org/hatewatch/2017/04/10/cross-burnings-still-'tool-fear'-used-racists

Sanchez, R., & O'Shea, K. (2016, December 10). Mass shooter Dylann Roof, with a laugh, confesses, 'I did it'. *CNN*. Accessed August 1, 2020, from www.cnn.com/2016/12/09/us/dylann-roof-trial-charleston-video/index.html

Schirch, L. (2005). *Ritual and symbol in peacebuilding*. Kumarian Press.

Schirch, L. (Ed.). (2016). *Handbook on human security: A civil-military-police curriculum*. Alliance for Peacebuilding, GPPAC, Kroc Institute.

Schirch, L. (Ed.). (2018). *The ecology of violent extremism*. Rowman and Littlefield.

Starr, D. (2002, July 29). Bad blood: The 9/11 blood-donation disaster. *The New Republic Online*. Accessed August 1, 2020, from www.academia.edu/386456/BAD_BLOOD_THE_9_11_BLOOD-DONATION_DISASTER_The_New_Republic

Sumiala, J. (2016, December 12). The politics of public mourning after terror attacks in France: How to grieve without fuelling anger? *Religion Going Public*. Accessed August 1, 2020, from http://religiongoingpublic.com/archive/2016/how-to-grieve-without-fuelling-anger-the-politics-of-public-mourning-after-terror-attacks-in-france

Lisa Schirch, PhD, is a senior fellow for the Toda Peace Institute, the Alliance for Peacebuilding, and George Mason University's Carter School for Peace and Conflict Resolution. She is the author of 11 books, including *Ritual and Symbol in Peacebuilding* (2004, Lynne Rienner Publishers). As a practitioner, Lisa has worked on institutional racism in the USA and has facilitated and participated in the design of national peace processes in Afghanistan and 12 other countries. *Blog:* lisaschirch.wordpress.com *E-mail:* lisa@toda.org

Nuclear Disaster, Trauma, and the Rituals of Scientific Method

Mae-Wan Ho, Alexey V. Nesterenko, Odile Gordon-Lennox, and Peter T. Saunders

Avant Propos

'The twentieth century brought us the bomb, and the nuclear threat will never leave us; the short-term threat from terrorism is high on the public and political agenda; inequalities in wealth and welfare get ever wider,' observes British cosmologist Lord (Martin) Rees (2003, p. 7). The risk of annihilation through nuclear war is compounded by poorly maintained nuclear reactors and the disposal of radioactive waste from nuclear power stations, which will remain

(continued)

This chapter is based on articles originally written by Dr. Mae-Wan Ho (1941–2016) and published by Science in Society. Editing includes updated information and data supplied by Dr. Alexey Nesterenko of IRS Belrad and by Odile Gordon-Lennox of IndpendentWHO.org, as well as a few definitions of scientific measurements and terms. Since Dr. Ho's untimely death in March 2016, Science in Society has ceased publication but the website remains available for consultation and the publications can be obtained from online suppliers. In light of the article's pertinence to education and dissemination of research about contemporary global threats, we are grateful to Dr. Peter Saunders and Dr. Eva Sirinathsinghji for permission to include it in this multiauthor volume.

M.-W. Ho
Institute of Science in Society, ISIS Foundation, London, Great Britain

A. V. Nesterenko
Institute of Radiation Safety (BELRAD), Minsk, Belarus
e-mail: anester@tut.by

O. Gordon-Lennox
Independent Researcher, Ferney-Voltaire, France

P. T. Saunders (✉)
King's College, Strand, London, Great Britain
e-mail: peter.saunders@kcl.ac.uk

© The Author(s), under exclusive license to Springer Nature Switzerland AG 2022
J. Gordon-Lennox (ed.), *Coping Rituals in Fearful Times*,
https://doi.org/10.1007/978-3-030-81534-9_12

toxic for many millennia. Recent nuclear disasters leave no doubt that national boundaries offer no protection. These disasters represent yet another ambiguous edge of our time. Lord Rees holds that the odds are against human survival to the end of this century.

This chapter describes the highly ritualised forms of scientific method used by scientists in the first hours of the Chernobyl disaster. At first, scientists prioritised diminishing the spread of radioactivity and radioactive fallout. Then, they began tackling how to reduce the human suffering caused by radioactive poisoning among contaminated populations. Mae-Wan was keenly aware of the high price exacted of scientists who practise the rituals of the scientific method with embodied intentionality in the ever-hostile environments around Chernobyl, and then Fukushima. Involvement requires them to rigorously respect those rituals but also to practise the rituals of publication, protest, and resistance. Some pay for their dedication with their reputation, others with their life, too many with both. Needless to say, their work is far from over. The search for practical solutions to the multi-faceted problems created by these accidents is ongoing.

Mae-Wan Ho did not mention the term 'ritual' in her original work on this chapter. If she were still with us, she'd have been able to write something that would explicitly mention the rituals of the scientific community. But of course that isn't possible at this stage. Mae-Wan Ho's scientific work was clearly grounded in her commitment to the application of her knowledge and research to social problems and undesirable, dehumanising conditions in contemporary society. This chapter honours Mae-Wan Ho and the scientists she defended with her pen, denounces the systemic edge revealed by the ongoing radioactive contamination no one wants to hear about, and decries the neglectful attitudes of political and economic powers that ignore the basic needs of the victims for truth and for adequate socio-medical structures.

Introduction

In the spring and summer of 1986 the Chernobyl power plant disaster released levels of radioactivity hundreds of times higher than the Hiroshima atomic bomb on hundreds of millions of people in the northern hemisphere, destroying normal life for tens of millions. Today, over thirty years later, more than six million people live on land with dangerous levels of contamination. Their children and their children's children are born there with no hope the situation will change over the next decades or even centuries.

Chernobyl fallout in contaminated territories continues to affect locally grown foodstuffs such as produce and cattle from farms and family gardens as well as wild game, fish, mushrooms, and berries. Parents who live in these areas know that it is dangerous to even drink the water. They want to be able to do something to help

Fig. 1 Protest against lack of protection for victims of the nuclear industry. A silent Hippocratic Vigil was held for 10 years outside the World Health Organisation headquarters in Geneva to remind the World Health Organisation of its responsibility to protect the health of people affected by the consequences of the nuclear industry. In 2008, Russian ecologist Alexey Yablokov (left) with Belarusian geneticist Rosa Goncharova (right) and Belarusian physicist Vassili Nesterenko (farthest right) joined to ritualise their protest in front of WHO headquarters in Geneva, Switzerland. On 26 April 2017, a commemorative stele was inaugurated at the site. | © Yann CC BY-SA

themselves and their children. They need answers to questions such as: How and where shall we live? How can we avoid the tragedy of bearing a child with malformations caused by irradiation? These are basic questions that first arose among the liquidators' families soon after the catastrophe. They remain, for the most part, unanswered (see Fig. 1).

As is the case in most areas contaminated by radioactivity, the majority of the programmes set up to help people are state-run or funded by the polluter. In the case of Chernobyl, this means that the major polluter measures and monitors the pollution, tries to clean it up, is responsible for preventing further damage, and cares for the affected population. In such a context is it imperative that independent entities monitor all foodstuffs as well as incorporated radionuclides in people and animals, determine individual cumulative doses using objective methods, and provide medical and genetic counselling, especially for families with children.

A Bit of Background

The Chernobyl nuclear disaster is widely considered to have been the worst nuclear accident in history and one of only two classified as a level 7 event on the International Nuclear Event Scale. The other is the Fukushima Daiichi nuclear meltdown, which took place on 11 March 2011 when Japan was hit by a magnitude 9 earthquake followed by a gigantic tsunami (Ho, 2011).

The Chernobyl accident occurred on 26 April 1986 at the Chernobyl Nuclear Power Plant near the city of Pripyat in Ukraine, then part of the Soviet Union, close to the border with the Republic of Belarus. A sudden power output surge prompted an attempt at emergency shutdown, but a more extreme spike in power output led to the rupture of a reactor vessel and a series of explosions. The graphite moderator was exposed, causing it to ignite. The resulting fire sent a plume of highly radioactive fallout over large parts of the western Soviet Union and Europe before winds blew it over North America and much of the northern hemisphere.

According to official post-Soviet data, about 60 per cent of the radioactive fallout landed in Belarus (Wikipedia, n.d.), contaminating vast areas with more than 37,000 becquerels[1] per m^2 (Bq/m^2). Between 1986 and 2000, 350,400 people were evacuated and resettled from the most contaminated areas of Belarus, Russia, and Ukraine. Agricultural production was halted on 264,000 hectares where two million people—of which 500,000 were children—lived (Nesterenko, 2001).

Champions of the Victims of Chernobyl

Vassili Nesterenko (1934–2008), a physician from Belarus and a former director of the Institute of Nuclear Energy at the National Academy of Sciences of Belarus, intervened personally during the accident at Chernobyl. In view of his expertise in nuclear energy and experience as a fire fighter, he decided to risk his life in the radioactive smoke in order to throw liquid nitrogen containers from a helicopter into the reactor core in an attempt to cool it. Although he survived for more than 20 years after the explosion, he eventually died as a result of the radiation he received at Chernobyl.

Nesterenko was one of the authors of a comprehensive report documenting the health impacts of Chernobyl (Ho, 2012a). Not only did this report cost Nesterenko his job at the National Academy of Sciences, it also resulted in his being threatened with internment in a psychiatric asylum. He escaped two attempts on his life. In 1989, with the help of Soviet physicist, dissident, and human rights activist Andrei Sakharov (Nobel Peace Prize winner in 1975), Belarusian writer and critic Ales Adamovich, and Russian chess grandmaster and former world champion Anatoly Karpov, Nesterenko created the independent Institute of Radiation Safety

---

[1] See the box *Scientific measurements and terms* at the end of this article.

'BELRAD'. Their goal was to study and document the consequences of the Chernobyl disaster. Nesterenko carried out radiation monitoring of the inhabitants of the zone contaminated by Chernobyl as well as their animals and foodstuffs. He developed instruments like the Human Radiation Spectrometer (HRS), an apparatus similar to a Geiger counter, for testing levels of contamination in humans. He served as director of BELRAD from 1990 until his death in 2008 when his son Alexey Nesterenko then succeeded him.

To date, BELRAD has performed 500,000 whole body count (WBC) measurements in 300 villages in the Belarusian provinces of Mogilyov, Brest, Grodno, Vitebsk, Minsk, and Bryansk. In 2001, the WBC laboratory of the Institute was officially accredited and certified. The scope of the work is large; it involves classification, evaluation, updating, and yearly publication of the all data collected by BELRAD in *The Radioecological Atlas: Human Beings and Radiation* (2001–2017). The Atlas represents a methodical analysis of the WBC measurements since 2001 of radiocaesium (Cs-137) levels monitored in children living in villages in 19 districts of the Chernobyl region of Belarus. WBC results from two additional provinces are included in the 2017 edition of the Atlas.

Vassili Nesterenko was not the only one persecuted for working on the consequences of Chernobyl. Two other scientists, Yuri Bandazhevsky, former director of the Medical Institute in Gomel (Belarus), and his wife Galina Bandazhevskaya,[2] a cardiac paediatrician, also dedicated their studies to understanding and mitigating the health consequences of the disaster. In June 2001, Bandazhevsky was sentenced to eight years imprisonment, as was the Deputy Director, Vladimir Ravkov. Their arrests came soon after Bandazhevsky published reports critical of the official research on the Chernobyl incident.

Bandazhevsky was released on parole from prison in 2005. Shortly afterwards, the mayor of Clermont-Ferrand in France, a city linked to Gomel since 1977, invited him to France where he collaborated with the French Commission de Recherche et d'Information Indépendantes sur la Radioactivité (CRIIRAD). Today, Bandazhevsky is director of the Ecology and Health Centre in Ukraine where he focuses on improving health conditions in the regions affected by the Chernobyl disaster. Dr. Bandazhevskaya continues her research on cardiovascular problems, in particular myocardial structural anomalies, observed among inhabitants of areas contaminated by the Chernobyl accident (Bandajevski et al., 2011; Bandajevski & Bandajevskaya, 2012).

---

[2] Alternative spellings of names written in Russian yield respectively: Bandazhevsky/Bandajevski and Bandazhevskaya/Bandajevskaya (see also the Reference section).

Chronic Absorption of Cs-137 by Children

While 90 per cent of children in Ukraine and Belarus were healthy in 1985, that figure dropped to 20 per cent after the Chernobyl disaster (Grodzinsky, 2009). All people living in the contaminated territories absorb radionuclides with food (up to 94 per cent), drinking water (up to 5 per cent), and through the air (about 1 per cent). On account of their lower weight and more active metabolism, children who eat the same food as adults accumulate up to five times more radionuclides. Moreover, the radionuclide levels of children living in rural areas are five to six times higher than city children of the same age.

For over 30 years, Bandazhevsky and Bandazhevskaya have documented the chronic assimilation of radiocaesium (Cs-137) into the organs of children living in contaminated areas (Bandazhevsky, 1998, 2000, 2001, 2003, 2012). There is evidence that the absorption of 50 Bq/kg of Cs-137 into a child's body can produce pathological changes in vital systems (cardiovascular, nervous, endocrine, and immune), as well as in the kidneys, liver, eyes, and other organs (Bandazhevskaya et al., 2004).

In the course of their work, they established that Cs-137 levels over 20 Bq/kg lead to disturbances in the electrophysiological processes of children's heart muscle. Children born after 1986 who have always lived in contaminated areas with concentrations above 15 Ci/km^2 (Ci, Curie = 3.7 x 10^{10} Bq) suffer serious pathological modifications of the cardiovascular system (Bandazhevskaya et al., 2004; Bandajevski et al., 2011; Bandajevski & Bandajevskaya, 2012; Ho, 2012a).

Between 1996 and 2000, between 70 and 90 per cent of the children living in contaminated areas where daily exposure to small amounts of radionuclides (mostly Cs-137) is virtually unavoidable had Cs-137 levels exceeding 15–20 Bq/kg. In many villages, levels reached 200–400 Bq/kg; the highest values were measured in Narovlya district with 6700–7300 Bq/kg. The highest accumulation was found in the endocrine glands, in particular the thyroid, the adrenals, and the pancreas. Elevated levels were also found in the heart, the thymus, and the spleen. This chronic accumulation of Cs-137 contributes to a progressive deterioration of the children's health (Bandazhevsky, 1998, 2000; Bandajevski et al., 2011; Bandajevski & Bandajevskaya, 2012).

How to Decrease Radionuclide Levels in Bodies

Early on, BELRAD determined that there are three basic ways to decrease the radionuclide levels in the bodies of people living in contaminated territories: reduce the amount of radionuclides in the food consumed, stimulate the body's immune and other protective systems, and accelerate removal of radionuclides from the body.

Parents and children can attend BELRAD training seminars where they can obtain the booklet 'How to Protect Yourself and Your Child from Radiation'.

The practical advice in the booklet includes how to reduce the levels of radionuclides in wild fowl, mushrooms, and fish by soaking them for two periods of 3–4 hours (2 tablespoons of salt with 1 tablespoon of vinegar in 1 litre of water) before cooking (Nesterenko et al., 2012). Soaking foods such as mushrooms in water and scalding, peeling, salting, or pickling vegetables also reduces the amount of radionuclides absorbed. Processing the fats found in foods such as milk and cheese can reduce also radionuclide levels several fold.

Certain vitamins and trace minerals can raise one's resistance to irradiation and stimulate the body's natural defences. Antioxidant vitamins like A and C as well as trace minerals such as Copper, Zinc, Selenium, Cobalt, and Iodine are known to inhibit the formation of free radicals. When added to the diet, these nutrients prevent the oxidation of organic substances caused by irradiation (lipid peroxidation). Sprouts of plants such as wheat, as well as seaweed (e.g., Spirulina), pine needles, mycelium, and other items can also increase resistance to the damage caused by irradiation (Yablokov et al., 2009 [2007]).

Apple Pectin for Reducing Radionuclides

In the search for ways to accelerate the removal of radionuclides from the body, Alexey Yablokov was the first to use a therapeutic form of apple pectin with children living in highly polluted areas of the Ukraine and eating contaminated food (see Fig. 2). He discovered that this pectin, then known as 'Yablopect', could reduce their Cs-37 load by 30 to 40 per cent. In Belarus, Vassili Nesterenko and his son Dr. Alexey Nesterenko then pioneered studies to test the efficacy of the treatment on 64 children from contaminated villages of the Gomel regions using dry, milled apple extract containing 15 to 16 per cent pectin in a randomised double-blind placebo-controlled trial (Nesterenko, 2001). During this study, conducted during a one-month stay in the Silver Spring sanatorium, all of the children ate only uncontaminated food. The average Cs-137 load in this group of children was about 30 Bq/kg body weight.

Results of the trials revealed a decrease in the Cs-137 burden in the bodies of all of the children. However, whereas the average decrease in those given a placebo powder or just unpolluted food was 13.9 per cent, Cs-137 levels in children taking the pectin powder were reduced by an average of 62 per cent. This means that the apple pectin treatment combined with a clean diet is 4.5 times more efficient than a clean diet alone. Of medical significance was the fact that, at the end of the month, none of the values for the children in the placebo group had dropped below 20 Bq/kg body weight. As noted above, Bandazhevskaya's studies showed that Cs-137 levels over 20 Bq/kg are associated with specific pathological tissue damage.

Furthermore, it is remarkable that no cases of intolerance or reduction in the children's vital nutrients were observed during the treatment. The researchers

Fig. 2 Vitapect: therapeutic grade apple pectin. The therapeutic grade apple pectin developed by BELRAD was made more effective with by adding certain vitamins and mineral nutrients. In collaboration with Jülich Research Centre in Germany, BELRAD created a joint database of the data stored at the two research institutes of all previous pectin treatments. Examination of the treatment results for 17,000 children from the age of 1 to 19 allowed them to determine the optimum dosage needed to reduce the Cs-137 burden in the body. When combined with clean food, the relative reduction of Cs-137 was 32% for the pectin groups compared with an average of 14% for the control groups. It is noteworthy that in the studies carried out by BELRAD, the total annual radiation dose for most of the children was below 1 mSv (the international limit of exposure). However, despite their relatively low doses, these children had wide-ranging health issues, which attests to the dangers of continuous, low-level radiation exposure. | Photo © S. Gordon-Lennox CC BY-NC-ND

concluded that this form of apple pectin was not only efficient but an extremely safe form of treatment that could be used even with young children (3 years and older).

Together, Bandazhevskaya and Nesterenko then studied cardiovascular health and Cs-137 decontamination by apple pectin treatment in a group of 94 children aged 7–17 years. The children received 5 grams of 16 per cent apple pectin powder twice a day with meals for 16 days. The study showed a reduction of the body burden of Cs-137 by 39 per cent from 38 to 29 Bq/kg and a rise from 72 per cent to 93 per cent in the number of normal EEGs. The fact that there was no improvement of cardiovascular symptoms suggests that, while pectin removes Cs-137, it does not necessarily correct the damage caused by irradiation (Bandazhevskaya et al., 2004).

Truth vs Indifference

In 2000, Kofi Annan, asked to write the preface to the Office for the Coordination of Humanitarian Affairs report on Chernobyl, concluded:

> The most vulnerable victims were, in fact, young children and babies, unborn at the moment when the reactor exploded. Their adulthood—now fast approaching—is likely to be blighted by that moment, as their childhood has been. Many will die prematurely. Are we to let them live and die, believing the world indifferent to their plight? (Annan, 2000)

Here is the shameful truth: If the government in Belarus were to agree that pectin is effective, they would have to acknowledge that there is widespread radioactive contamination in the country. Then they would have to admit that they (and the West) have allowed their people, two million, including 500,000 children, to eat contaminated food for 30 years, to become ill, live miserable lives, and die premature deaths. Moreover, had it not been for a scandalous disinformation campaign mounted against the apple pectin treatment, which stopped major funding from the European Parliament in the 1990s, BELRAD would have made much more progress (Tchertkoff, 2006; Greaves, 2012).

Seaweed Alginate for Radioprotection

Radioprotection is an urgent issue, not only for the victims of Chernobyl but also now for those living in highly contaminated areas around Fukushima (Ho, 2012b). A study carried out at the Institute of Radiation Medicine in Beijing, China in 1991 demonstrated that sodium alginate prepared from seaweeds such as *Sargassum* sp. and kelp (*Laminaria* sp.) was able to block radioactive strontium uptake (Gong et al., 1991). Sodium alginate from *S. siliquastrum* in particular, reduced the body burden of strontium 3.3–4.2 fold in rats, and by 78 per cent (+/− 8.9) in human subjects. No undesirable effects on gastrointestinal function were observed, nor were Calcium, Iron, Copper, and Zinc metabolism altered, either in the animal experiments or in human volunteers. In a more recent study at the Institute of Radiation Protection in Ingolstädter, Germany, researchers found that sodium alginate added to Sr-90 contaminated milk reduced the uptake of Sr-90 by a factor of 9 (Hollriegl et al., 2004). The seaweed nori in the Japanese diet is also a rich source of alginate.

Worldwide Ramifications for the Nuclear Industry

The ramifications for the nuclear industry worldwide need hardly be stated (Greaves, 2012). Chernobyl and Fukushima are just two among many similar disasters waiting to happen worldwide. There are currently 447 operable civil nuclear power reactors around the world, with a further 61 under construction.[3]

Radionuclides are found in the air, water, forests, dust, grass, and food around nuclear power plants and uranium mines. They settle in the organs of humans and animals. The effect of the daily emission of radioactivity on the health of those living near a nuclear plant is well known (Körblein & Hoffman, 1999; Scherb et al., 2016a, 2016b) but governments and regulators systematically downplay the risks and hide the real costs of nuclear power and mining. Studies show a decline in female births among exposed fathers as well as an increase in hereditary congenital conditions such as Down's Syndrome and cancer among children whose parents were subjected to low-dose exposures while living near these power plants before their child's birth (Schmitz-Feuerhake, 2014). There is no place for nuclear in a truly green energy portfolio and there is a lot we can still do.

Hope

Scientists like Galina Bandazhevskaya, Yuri Bandazhevsky, Vassili Nesterenko, Alexey Yablokov, and Alexey Nesterenko dared to imagine the far-reaching consequences of a nuclear accident of mindboggling proportions. Not only did they respect the spirit of the Hippocratic Oath in abstaining from all intentional wrongdoing and harm, but they searched for and promoted cures to help the sick, without administering poison. While their work has made a real difference in the lives of the villagers who have seen their levels of radionuclides reduced, the radiological contamination of people in other places remains the same, or even worsens.

As public resources and energy are expended to try to keep nuclear disasters invisible, women and men feel compelled to do something. In the face of official indifference, they respond creatively—often at great professional and personal cost—to the suffering of the victims with the means they have to hand such as their scientific method, their writing, or social protest (see Fig. 1).

Admitting to the errors of the past is a first step towards putting the nuclear genie back into the bottle. The importance of decontamination, continuing health surveillance, and radioprotection for the victims of Chernobyl and Fukushima cannot be over-emphasised. Developing more non-toxic cures to alleviate the suffering of the victims of nuclear radiation is yet another step that can be taken. As long as we foster this kind of bold and innovative work there is indeed hope for future generations to recover health and vitality. They deserve all our support.

---

[3] A list of reactors operable, under construction, planned, and proposed is updated monthly (WNA, 2018, 2019).

For More Information

Outside funding is essential for the work at BELRAD in Belarus and at Ecology and Health Centre in Ukraine. If you would like to help, please contact the following organisations.

Institute of Radiation Safety 'BELRAD'

website: http://belrad-institute.org
e-mail: irs.belrad@gmail.com

Ecology and Health Centre

e-mail: center.by.ukr@gmail.com

Other Organisations

- IndependentWHO.org
- International Physicians for the Prevention of Nuclear War: www.ippnw.org
- Enfants de Tchernobyl Belarus: enfants-tchernobyl-belarus.org

Simulated Mapping of the Spread of the Disasters

- Chernobyl radioactive dispersion over Europe and Asia (animation by CRNS, France). Accessed on 5 August 2020 at www.youtube.com/watch?v=Fw02i7iW-OQ
- Fukushima radioactive aerosol dispersion over the northern hemisphere (animation by NASA, USA). Accessed on 5 August 2020 at www.youtube.com/watch?v=HCzuPm4T4qo

Scientific Measurements and Terms

- **Alginates** are manufactured from brown seaweed of the phylum Phaeophyceae. They reduce the body burden of radioactive strontium (Sr-90) by blocking uptake of this heavy metal.
- **Becquerel** (Bq), **Curie** (Ci), **Sievert** (Sv): These three names refer to units of measure used in the International System of Units.

  The becquerel (Bq) is a unit used for measuring radioactivity, more precisely, the activity of a quantity of radioactive material in which one nucleus decays per second. The becquerel was named after Henri Becquerel. It succeeded the curie (Ci), an older unit used to measure quantities of radioactive material named after Pierre and Marie Curie. In 1903, the three scientists shared a Nobel Prize in Physics for their work in discovering radioactivity.

  The sievert (or millisievert, abbreviated mSv), a derived unit of ionising radiation dose, is a measure of the health effect of low levels of ionising radiation on the human body. The sievert is of importance in dosimetry and radiation protection. It is named after Rolf Maximilian Sievert, a Swedish medical physicist renowned for work on radiation dose measurement and research into the biological effects of radiation.
- **Free-radicals**: Gerhard Herzberg, who won the Nobel Prize for his research into the electron structure and geometry of radicals, suggested a loose definition of free-radicals adopted by many researchers: 'any transient (chemically unstable) species (atom, molecule, or ion)' (1971).
- **Radionuclides** are atoms with excess nuclear energy that makes them unstable, and eventually decay. The decay rate of a radioactive isotope is measured in half-life. Half-life of a radioactive substance is defined as the time required for disintegration of one half of the original amount of the radioactive isotope.

  Although low levels of exposure occur naturally without harm, unplanned exposure to radionuclides generally has a harmful effect on living organisms, including humans. They contribute to the deterioration of the health of people exposed to the open-air atomic bomb testing in the 1950s, the wide use of impoverished uranium in armaments and military equipment, the Chernobyl disaster, and the Fukushima nuclear crisis.

  - **Cs-137**: Caesium-137, or radiocaesium, is a radioactive isotope of caesium. It is among the most problematic of the short- to medium-lifetime fission products because it moves and spreads easily in nature due to the high water solubility of caesium's most common chemical compounds, which are salts.

(continued)

- **I-131**: Iodine-131 is a short-lived radioisotope of iodine that has a radioactive decay half-life of about eight days. It settles in the thyroid gland and may cause cancer.
- **Other radioactive isotopes**: Independent studies draw attention to the dangers of chronic ingestion of small doses of long-lived radioactive particles such as C-137, which has a half-life of 30 years, as well as of Strontium (Sr-90) with a half-life of 29 years and Plutonium (Pu-239) with a half-life of 24,100 years.

References

Annan, K. (2000). Report preface. *Chernobyl: A continuing catastrophe*. United Nations Office for the Coordination of Humanitarian Affairs, OCHA/99/20.

Bandazhevskaya, G. S., Nesterenko, V. B., Babenko, V. I., Babenko, I. V., Yerkovich, T. V., & Bandazhevsky, Y. I. (2004). Relationship between caesium (Cs-137) load, cardiovascular symptoms, and source of food in 'Chernobyl' children: Preliminary observations after intake of oral apple pectin. *Swiss Medical Weekly, 134*, 725–729.

Bandazhevsky, Y. I. (1998). *Pathophysiology of incorporated radioactive emission*. Gomel State Medical Institute.

Bandazhevsky, Y. I. (2000). *Medical and biological effects of radiocaesium incorporated into the human organism*. Minsk: The Institute of Radiation Safety 'BELRAD'. Accessed August 5, 2020, from https://pdfs.semanticscholar.org/eb51/b0259476a7b82857bc78507180a20c66c4da.pdf

Bandazhevsky, Y. I. (2001). *Radioactive caesium and the heart: Pathophysiological aspects*. Minsk: The Institute of Radiation Safety 'BELRAD'. Accessed August 5, 2020, from www.ratical.org/radiation/radioactivity/RadCsAndHeart2001.pdf

Bandazhevsky, Y. I. (2003). Chronic Cs-137 incorporation in children's organs. *Swiss Medical Weekly, 133*, 488–490. Accessed August 5, 2020, from https://pdfs.semanticscholar.org/f9f3/edd723387e281c20dd5152d74ed1c32e39ad.pdf

Bandazhevsky, Y. I. (2012). *From the syndrome of chronic incorporation of long lived radionuclides (SLIR) to the creation of programmes and radioprotection policies for populations, an example of an integrated model*. Presented at the Scientific and Citizen Forum on Radioprotection – From Chernobyl to Fukushima, 11–13 May, Geneva, Switzerland. Accessed August 5, 2020, from http://independentwho.org/en/books

Bandajevski, Y. I., Bandajevskaya, G. S., & Dubovaya, N. (2011). *Tchernobyl, 25 ans après: Situation démographique et problèmes de santé dans les territoires contaminés*. Yves Michel Éditions.

Bandajevski, Y. I., & Bandajevskaya, G. S. (2012). *Les conséquences de Tchernobyl sur la santé: Le système cardiovasculaire et l'incorporation de radionucléides Cs-137*. Yves Michel Éditions.

BELRAD. (2001–2017). *The radioecological atlas: Human beings and radiation*. Minsk: BELRAD. (The electronic version in Russian, updated quarterly, is available upon request. Digital copies are stored in the National Library of the Republic of Belarus.)

Gong, Y. F., Huan, Z. J., Qiang, M. Y., Lan, F. X., Bai, G. A., Mao, Y. X., Ma, X. P., & Zhang, F. G. (1991). Suppression of radioactive strontium absorption by sodium alginate in animals and human subjects. *Biomedical and Environmental Sciences, 4*, 273–282.

Greaves, S. (2012). The pectin controversy. *Science in Society, 55*. Accessed August 5, 2020, from www.i-sis.org.uk/The_Pectin_Controversy.php

Grodzinsky, D. M. (2009). Foreword. In Yablokov et al. (Eds.), *Chernobyl: Consequences of the catastrophe for people and the environment* (pp. vii–xi). Blackwell Publishing.

Herzberg, G. (1971). *The spectra and structures of simple free-radicals: An introduction to molecular spectroscopy*. Cornell University Press.

Ho, M. W. (2011). Fukushima nuclear disaster. *Science in Society, 50*. Accessed August 5, 2020, from www.i-sis.org.uk/Fukushima_Nuclear_Disaster.php

Ho, M. W. (2012a). Chernobyl deaths top a million based on real evidence. *Science in Society, 55*. Accessed August 5, 2020, from www.i-sis.org.uk/Chernobyl_Deaths_Top_a_Million.php

Ho, M. W. (2012b). Truth about Fukushima. *Science in Society, 55*. Accessed August 5, 2020, from www.i-sis.org.uk/Truth_About_Fukushima.php

Hollriegl, V., Rohmuss, M., Oeh, U., & Roth, P. (2004). Strontium biokinetics in humans: Influence of alginate on the uptake of ingested strontium. *Health Physics, 86*, 193–196.

Körblein, A., & Hoffman, W. (1999). Childhood cancer in the vicinity of German nuclear power plants. *Medicine and Global Survival, 6*(1), 18–23. Accessed August 5, 2020, from www.ippnw.org/pdf/mgs/6-1-korblein.pdf

Nesterenko, V. B. (2001). Radioprotective measures for the Belarusian population after the Chernobyl accident. *International Journal of Radiation Medicine, 3*, 12.

Nesterenko, V. B., Nesterenko, A. V., Babenko, V. I., Kozyrenko, M. A., Krasnopyorov, I. V., & Voida, O. A. (2012). *Implementation of radioprotection for populations at local level. Radio-ecological atlas: Human beings and radiation*. Presented by Alexey Nesterenko at the Scientific and Citizen Forum on Radioprotection – From Chernobyl to Fukushima, 11–13 May, Geneva, Switzerland. Accessed August 5, 2020, from http://independentwho.org/en/books

Rees, M. J. (2003). *Our final hour: A scientist's warning: How terror, error, and environmental disaster threaten humankind's future in this century – on Earth and beyond*. Basic Books.

Scherb, H., Kusmierz, R., Sigler, M., & Voigt, K. (2016a). Modelling human genetic radiation risks around nuclear facilities in Germany and five neighbouring countries: A sex ratio study. *Environmental Modelling & Software, 79*, 343–353. Accessed August 5, 2020, from www.sciencedirect.com/science/article/pii/S1364815215300773

Scherb, H., Kusmierz, R., & Voigt, K. (2016b). Human sex ratio at birth and residential proximity to nuclear facilities in France. *Reproductive Toxicology, 60*, 104–111. Accessed August 5, 2020, from https://doi.org/10.1016/j.reprotox.2016.02.008.

Schmitz-Feuerhake, I. (2014). *Immediate and delayed genetic effects of ionizing radiation through irradiation and contamination*. Presentation at the Scientific and Citizen Forum on the Genetic Effects of Ionizing Radiation, 29 November, Geneva, Switzerland. Accessed August 5, 2020, from www.lulu.com/shop/independentwho/proceedings-of-the-scientific-and-citizen-forum-on-the-genetic-effects-of-ionizing-radiatio/paperback/product-22427584.html

Tchertkoff, W. (2006). *Le Crime de Tchernobyl ou le goulag nucléaire*. Actes Sud.

Wikipedia. (n.d.). *Chernobyl disaster*. Accessed August 5, 2020, from https://en.wikipedia.org/wiki/Chernobyl_disaster

World Nuclear Association (WNA). (2018). *Number of nuclear reactors operable and under construction*. Accessed August 5, 2020, from www.world-nuclear.org/nuclear-basics/global-number-of-nuclear-reactors.aspx

World Nuclear Association (WNA). (2019). *World nuclear power reactors and uranium require-ments*. Accessed August 5, 2020, from www.world-nuclear.org/information-library/facts-and-figures/world-nuclear-power-reactors-and-uranium-requireme.aspx

Yablokov, A. V., Nesterenko, V. B., & Nesterenko, A. V. (2009 [2007]). *Chernobyl: consequences of the catastrophe for people and the environment*. Blackwell Publishing.

Mae-Wan Ho (1941–2016), PhD, was a scientist who made significant contributions to many fields, from molecular genetics and evolution to the physics of organisms and indeed physics itself. She wrote nearly 200 scientific papers. Her best-known monographs are *The Rainbow and the Worm* (1993, now in its 3rd edition) and *Living Rainbow H₂O* (2012), both published by World Scientific. In 2014, she received the Prigogine Medal for her work.

Mae-Wan founded the Institute of Science in Society (1999) and both directed it and edited its magazine, *Science in Society*, for 17 years. She wrote over 700 articles as well as several books and pamphlets. These were popular works, in the sense that they were written to be accessible to the layperson, but they were scientifically rigorous and fully referenced. Many are still available on the *ISIS archive*: i-sis.org.uk

Alexey V. Nesterenko, PhD in ecological science, is director of the independent Institute of Radiation Safety (BELRAD) in Minsk, Belarus He has written more than 40 publications on the environment and on radioprotection. In 2007, in collaboration with his father, biologist Vassily B. Nesterenko (1934–2008), and biologist Alexey V. Yablokov, he wrote a landmark work in Russian, known in English as *Chernobyl: Consequences of the Catastrophe for People and the Environment* (2009, Blackwell). *Website:* belrad-institute.org *E-mail:* anester@tut.by

Odile Gordon-Lennox holds degrees in law and political science from the University of Grenoble in France. She first became aware of the dangers of uranium mining through the early investigations of her mother-in-law, activist Diana Kingsmill Wright, on the effects of radioactivity on native Canadians living around mines in northern Saskatchewan in the 1950s. Uranium mines in some 20 countries still poison rivers and forests, generate radioactive dust, and produce ore waste, which is often used as landfill in public areas. As member of the Independent WHO Collective, Odile actively supports the French NGO Enfants de Tchernobyl Belarus.

Peter Saunders, PhD, is Emeritus Professor of Mathematics at King's College, London. He was co-director of the Institute of Science in Society and deputy editor Science in Society. He has written more than 100 scientific papers and also many articles for Science in Society on various topics including nuclear energy. *Science in Society Archive Website:* https://www.i-sis.org.uk

'Dead Land Dead Water'

Nowhere Left to Go

Jeltje Gordon-Lennox

Once upon a time, our planet had pristine air, fertile land, and pure water. Animals and humans moved about freely, leaving few traces in their wake. The earth is still alive. But there is much to indicate that we live and love on borrowed time. Our planet's capacity to support us is increasingly compromised by what sociologist Saskia Sassen identifies as 'a global multisited array of dead patches of land and water in the tissue of the biosphere' (2014, p. 150). Despite general consensus about the remedy for the earth's condition—initiate creative changes with respect to humans and nature—there is far less agreement on exactly what should be done, and who should pay for it. In the face of this threatening diagnosis, the resulting stalemate begs the question: What happens when it *is* too late?

This chapter explores the role of ritualising at what Sassen refers to as the 'systemic edge'.[1] The cases presented concern crises that occurred suddenly or unexpectedly with little leeway for defence or escape. They document, in bits and pieces, an overarching dynamic that exposes new phases of global capitalisms. Left

---

The phrase 'Dead Land Dead Water' wittingly ties this chapter to Saskia Sassen's landmark work *Expulsions: Brutality and Complexity in the Global Economy*. Expulsion, Sassen observes, is no longer intrinsically linked to poverty or inequality, or even to vaccination, but to global phenomena. Now, once expelled from their livelihoods, homes, or the biosphere that makes life possible, people may be prodded towards a systemic edge where they become invisible.

This is a personal article in the sense that I know many of the people in the cases presented. Some are composite characters; some of the names and circumstances have been changed. My family and I survived one of the disasters described.

---

[1] The 'systemic edge' is an organising assumption in Sassen's book *Expulsions*. It is a place 'where general conditions take extreme forms precisely because it is the site for expulsion or incorporation. Further, the extreme character of conditions at the edge makes visible larger trends that are less extreme and hence more difficult to capture' (2014, p. 211).

J. Gordon-Lennox (✉)
Psychotherapist ASP, Geneva, Switzerland
e-mail: jeltje@gordon-lennox.ch

with a sense that the disaster could have been prevented, the central figure(s) feel overwhelmed and isolated by the scale and the ongoing nature of the crisis. Sooner or later they will face death or expulsion, with nowhere left to go.

Threat, Fear, and Trauma

Threat, fear, and traumatic events are part of everyday life for animals and human beings alike. Even so, not all creatures are traumatised. When events, such as the birth of a child, that first kiss, an achievement, or a disaster divide time into 'before' and 'after', we tend to remember the year, the day, and perhaps that split second when our life changed. The images and sensations of the experiences that touch our lives in this way are fixed, first in our bodies and then in our memories. When trauma occurs at a very young, pre-verbal age, the experience is stored in the body, with little or no memory. French psychiatrist Pierre Janet (1859–1947) observed that traumatic memories may become fixed ideas that not only alternate with our habitual personality but also intrude upon it, particularly when we are confronted with salient reminders of the trauma.

When threatened, wild animals fare best when they can react with fight or flight. Humans and domesticated animals too experience less long-term trauma when they can defend themselves or flee. Healthy responses to threat among wild animals are increasingly thwarted as their wilderness habitats dwindle. The same applies to humans and domesticated animals subjected to relentless dislocation and overcrowding. These conditions create intense feelings of helplessness and contribute to making us a traumatised species. [2] In turn, the physiological and psychological effects of not feeling safe—even in our own bodies—foster fragmentation in all our relationships.

Feeling safe goes beyond the removal of threat. The fundamental human quest for a sense of safety begins at birth 'when the infant's needs for soothing are dependent on the mother. This quest for safety through co-regulation continues throughout the lifespan', compels us to search for creative ways to calm our neural defences, and motivates us to create 'trusting friendships and loving partnerships' (Porges, 2015, p. 122). Ritualising is one of the ways we have developed to sooth ourselves individually and collectively (Malinowski, 1948 [1925]) (see Fig. 1).

---

[2] During the Leningrad Floods in the 1920s, a number of the caged dogs in Russian physiologist Ivan Pavlov's (1849–1936) basement laboratory drowned. He found that, after experiencing the tremendous inescapable stress of near drowning, the surviving dogs' behaviour changed, e.g. dogs that had been affectionate no longer trusted him. Most important for our understanding of human reactions to high stress is Pavlov's discovery that he could wear a dog down by subjecting it to excessive work (on a treadmill), upsetting its stomach with bad food or irregular feedings, or inducing a fever (Time Magazine, 1957).

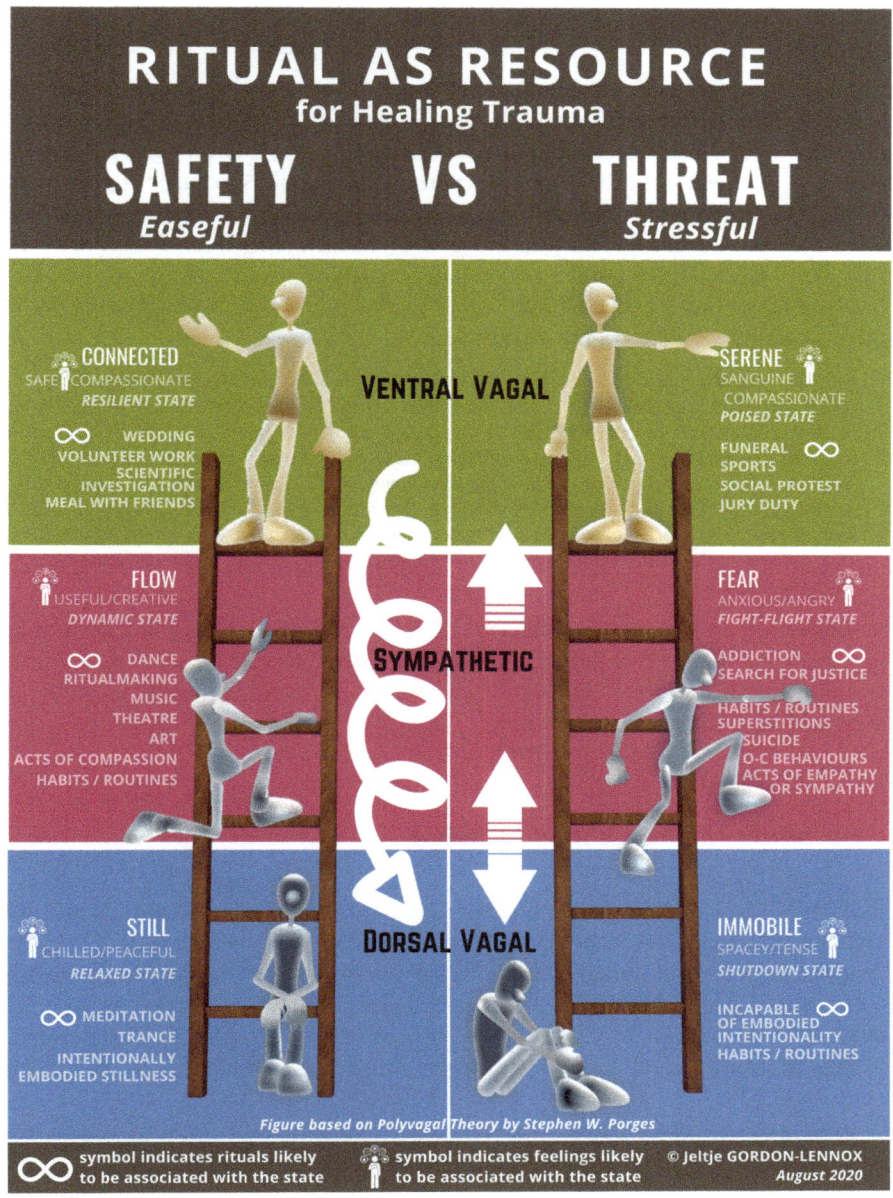

Fig. 1 Ritual as resource for healing trauma. Ritual practice is an adaptive mechanism for coping with ongoing threat of disaster and a resource for healing trauma. Ritual is not therapy, but it can indeed be therapeutic. 'A careful investigation of many rituals results in the discovery that the rituals are functional exercises of vagal pathways' (cf. Porges' chapter, Part I). Like neural exercise, ritualising impacts *how* we are. This suggests that enacting specific religious, secular, or routine elements is less important than what I call embodied intentionality. This term refers to a subjective experience that involves 'gut-up' simultaneous dual awareness of the ritual experience and of one's own bodily sensations. Such intentionality may well be what distinguishes ritual activity from non-ritual practices such as ordinary habits, routines, obsessive-compulsive behaviours, and addiction (see also Introduction). | © J. Gordon-Lennox

Ritual: An Adaptive Coping Mechanism

From the beginning of time, humankind appears to have countered the fear and anxiety of illness, transience, and mortality by making and practising ritual. Creative practices around burial represent some of the earliest rituals known to us. After the attacks of 11 September 2001 in New York, many people spontaneously gathered near 'ground zero' or in other public places to engage in artistic expressions of their grief and fear. Similar public ritualising occurred after attacks in European cities. Ethologist Ellen Dissanayake, who refers to ritualising as 'making special' and 'making the ordinary extraordinary', qualifies this profound emotional and visceral drive to act an adaptive coping mechanism:

> Art-filled ritual practices address and satisfy evolved needs of human psychology. They create and reinforce emotionally reassuring and psychologically necessary feelings of close relationship with others and of belonging to a group. Further, they provide to individuals a sense of meaningfulness or cognitive order and individual competence insofar as they give emotional force to explanations of how the world came to be as it is and what is required to maintain it. They are adaptive not only because they join people together in common cause but because they also relieve anxiety. It is better to have something to do, with others, in times of uncertainty rather than try to cope by oneself or do nothing at all. (Dissanayake, personal communication, 2016)

Embodied art-filled ritual, whether traditional or emerging, can initiate healing processes and mitigate the negative impact of a traumatic event by soothing our fears and drawing us together. [3] The three cases below explore how people use ritual to cope with extreme ongoing threat.

Case Study 1: Flash Flood in Nepal

On 5 May 2012 at around nine o'clock in the morning in north-central Nepal, the collapse of a ridge—probably initially glacier ice—near Annapurna IV (7526 m) dropped thousands of tonnes of ice and rock almost vertically onto unstable debris composed of glacial moraines, ancient glacial lake silts, and gravels that rested in the deep bowl of the Sabche Cirque some 3000 m below (see Fig. 2). The powerful avalanche caused earth tremors measuring from 3.8 to 4 on the Richter Scale

---

[3] Despite Pierre Janet's pioneer work (1859–1947), trauma is still too often reduced to a mental condition for which sufferers commonly receive talk, cognitive-behavioural, re-exposure, and drug therapies. Recent research in neurophysiology offers alternative theories that open the way for body-based treatments. The efficacy of ritual in healing trauma involves all of the various strata of the brain, and more specifically the ventral vagal social engagement theory (Porges, 2011), which addresses the functions of the brain stem, the mid-brain, and the left orbito-frontal cortex – virtually all of the layers of the brain, and the right more than the left cortex (Scaer, personal communication, 2016).

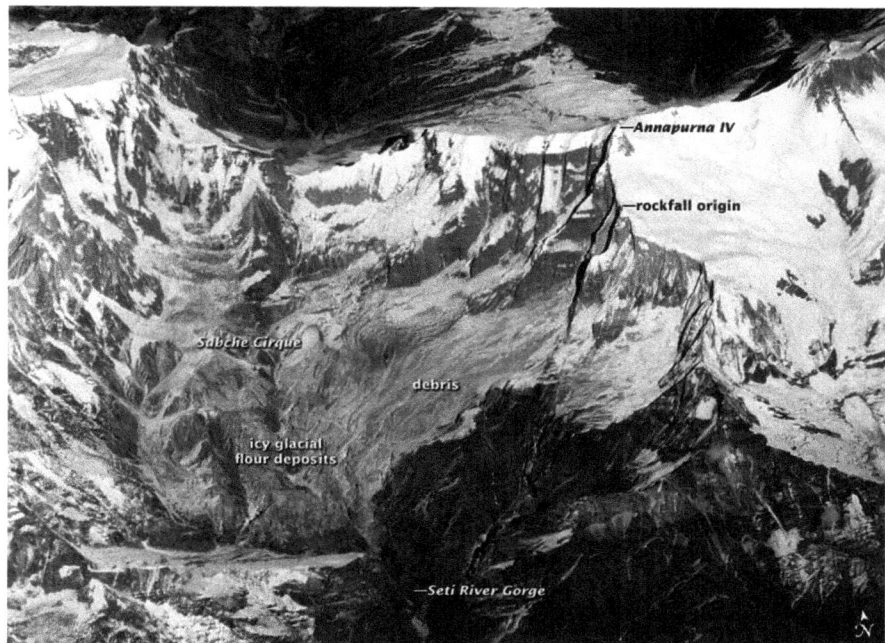

Fig. 2 A chain of events caused the Seti Khola disaster. It took experts and local residents nearly two years to lift the mystery on what had caused the flash flood. The total drop from the ridgeline (rockfall origin) above Sabche Cirque to the river bed of the Seti Khola below the city of Pokhara is about 6100 m spread over a distance of only 40 km. The event caused enormous erosion, depositing debris all along the river basin and channelling it into the Pokhara Valley, Nepal's second largest concentration of population. | © Earth Observatory NASA Public Domain

(SANDRP, 2014). It gathered up some of the loose debris and propelled the mass another 1500 m lower into the gorge of the Seti Khola (White River).

The grey hyperconcentrated slurry released on that fateful Saturday morning came in 27 waves contained huge ice blocks that raced at more than 30 m per second through the narrow Seti Khola basin. At half past nine, a flash flood hit Khara Pani, a small riverside settlement in the Kaski District of the Annapurna region, sweeping up picnickers, bathers at the thermal springs, and people working in the riverbed, as well as livestock and homes. Some 72 people including 3 tourists were killed.

At the three-year commemoration of the Seti Khola disaster, Sumila Darji recounts:

When the flood hit, my husband's mother, his brother and his nephew were in the riverbed. Sarila [sister-in-law] and I watched helplessly as that 'grey snake' overtook them and many goats.

Everyone who has electricity listens for the radio warnings about avalanches, floods—and climate change. Landslides are common in Nepal, especially where new roads and bridges are built. Before, the winters were colder and the rains would come only after the monsoon rituals at the *gompa* [monastery]. Now the rains come whenever they wish. Some of the smaller streams are disappearing but our river

(continued)

swells. That Saturday was different. No clouds, no rain; the flood came out of nowhere. . . . How were we to know?

Before the flood, my husband, his mother and I lived with our two children near Khara Pani at a big curve in the river, not far from the footbridge. Now, Sarila and her baby live with us. We are tailors, musicians and day-labourers. Once, my husband went foreign [abroad] but a big metal beam fell on his leg. When it is cold and rainy his leg hurts. Since the flood he drinks more.

Sarila's eyes are empty, sometimes wild. She often cries out in her sleep but still does not speak. She constantly splashes her face with cold water. When we massage our babies, Sarila's breathing changes; sometimes she hums and smiles at the baby.

We protect our goats now with more red ribbons from Muktinah [the local sanctuary] and more bells. We listen for the bells and chase the goats up the hill when they go too near the river. We also paint special signs on the riverside door [see Fig. 3] and make offerings [flowers, rice, oil candles] at all the temples. Even so, maybe, one day the river will take our hut, and us too.

Residents recalled that the Seti's normally turbid white rapids or 'glacial milk' had momentarily turned yellow; at the time, they put this down to bridgework upstream. Then the river slowed to a mere trickle of abnormally clear blue water. These were the only signs of impending danger.

In the end, it was concluded that a chain of events triggered the disaster. An earlier series of small rockfalls had obstructed the outlet to a previously unknown and unimaginably vast natural reservoir hidden deep in the Seti Khola gorge (Kargel, 2014). The forceful avalanche, which had shaken the earth and flattened entire forests, unblocked the mouth of the gorge, sending the deadly slurry racing down the chute created by the narrow valley towards Pokhara.

From Villagers to Displaced People

About ten thousand years ago, Pokhara Valley was covered by a natural hazard caused by melting ice, avalanches, and hidden reservoirs. What happened in 2012 was no 'natural disaster'. There's actually 'no such thing as a natural disaster', affirms Kendra Pierre-Louis. 'Hazards are natural; disasters are manmade' (2017). Disasters happen when there is risk and loss of life. If Sumila should move—or be relocated—with her family to higher ground, they would cease being in danger. At least statistically. In reality, the Darji family would just go from one systemic edge to another on an increasingly unstable mountainside.

Rapid changes in temperature, population growth, and frenetic road building is a dangerous combination for environmental stability and human safety. 'Something like this will happen again', says Kargel, 'it's inevitable' (2014). A new catastrophe on this or any of the other rivers flowing down from the Himalayan Mountains could have an apocalyptic impact on the area. Today, about a half a million people live in the Pokhara Valley.

Ritualising in Fearful Times: Natural Hazards

In analysing responses to two major earthquakes in Nepal—one in 1934 and another in 2015—disaster prevention researcher Roshan Bhakta Bhandari affirms that ritual practices in the Kathmandu Valley enabled people to reorganise their normal lives

Fig. 3 Signs of protection. This door bears four signs of protection: ochre paint, two prints of a right hand on each side of the opening, a motif above the entrance, and a padlock. | © John Pavelka CC BY

and cope with the social uncertainties caused by these disasters.[4] In 1934, the residents' involvement (organisation and participation) in traditional ritual events definitely enhanced their community's capacity to cope with the risk of disaster, and perhaps even the disaster itself (Bhandari et al., 2010, 2011; Bhandari, 2014).

In the interval between the two earthquakes, Nepalese society changed significantly. The most remarkable difference is the growth and urbanisation of the population. Second, people now have access to outside sources of information (radio, internet), as well as different needs, expectations, and resources. Despite these changes, in 2015 as in 1934, traditional and ritualised practices contributed to survival by meeting people's physical, emotional, and spiritual needs to feel safe in proximity to others (self-regulation), and to engage with others (co-regulation), thus strengthening social bonds (see Fig. 1). Bhandari observes:

> People got together with their neighbours in open spaces and spent several nights in temporary shelters. As they shared their food and their feelings, they transformed some of the trauma of the tragedy. Those who lost family members were able to communicate their grief to their close neighbours who took the time to console them. The socio-psychological significance of these communal practices was invaluable in coping with the immediate and long-term stress of losing loved ones and property. (Bhandari, personal communication, 2019)

---

[4] In the case of the 1934 earthquake in Lalitpur, near Kathmandu, Bhandari's research involved an official study and interviews of survivors (2010, 2011, 2014). His perspective on the 2015 earthquake is based on first-hand accounts of family and friends near Kathmandu who survived the disaster (personal communication, 2019).

Gender and age influences how people cope with threat and disaster: the most vulnerable[5] are most likely to seek out and practise ritual activities. In rural Khara Pani, official commemorations represent the most obvious occasions for co-regulation among the survivors of the Seti Khola flash flood. The villagers also practise family rituals to mark their losses, such as maintaining altars with photos and objects that belonged to the dead.

Sumila makes ordinary objects and behaviours extraordinary through intentional private rituals (ribbons and bells on the goats, protective symbols on children and doorways) that are designed to help her and her family feel safe (probably performed in a ventral vagal or sympathetic state). Sumila's husband's addictive behaviour may be an adaptive mechanism (sympathetic state) for dealing with the disaster and his earlier injury. Sarila's mutism is a sign of her limited sense of safety (dorsal vagal state) (see Fig. 1). Her responsiveness to physical touch (cold water, massage) could be a first step towards her reincorporation in the family and larger community. Whether or not these behaviours are rituals depends on the embodied intentionality of those who perform them.

Case Study 2: Hazardous Agrochemicals in Brazil

The famous Iguaçu Falls, located on the border of Argentina and Brazil, used to be surrounded by forest and grasslands with small hardy trees, thick brush, and smaller wild animals like jaguars, capybara, and howler monkeys. Now, this wildlife is rarely found outside of local zoos.

> As far as the eye can see, the southern Brazilian state of Paraná is covered with fields of soya or maize bearing signboards with the names of international agrochemical companies.
>
> 'I've been working for 10 years at the Fazenda Perfeita,' observes Vagner, a 32-year-old agricultural worker at a foreign-owned industrial farm located near Cascavel in Paraná. His 24-year-old common-law wife Marilea cleans and cooks for the family of one of the local administrators of the fazenda. 'We met two years earlier at a Christmas dance in a small town nearby,' recounts Marilea. 'Our son Diego, born at term on 28 February 2020 in the local hospital, was pronounced healthy. Three weeks later he stopped nursing and turned blue. We rushed him to hospital where we learned that he had multiple heart malformations.'
>
> 'After many tests,' adds Vagner, 'the doctors told us that Diego might not live long enough to learn to crawl or walk. He needs a heart operation that can only be done in Curitiba [the capital of Paraná, a seven-hour bus ride away] or in São Paulo [14 hours away by bus].' Marilea stopped working to care for him at home. 'A nurse

(continued)

---

[5] Women, girls, the aged, and young children (Evetts, 2017) are particularly vulnerable. In rural areas of Nepal, women are responsible for collecting firewood and water. Shortages increase their workload and walking distance, which impacts their health, food security, personal hygiene, and mortality (e.g. drowning during a flood) is higher among women and girls (Dhimal, 2015).

we know admits she sees more and more babies like Diego from families living on or near the big farms where pesticide use is part of daily life.' Marilea angrily adds: 'Why do they let this happen?'

The couple do not have the means to bring their child for treatment at a major hospital. Marilea practises cures learned from her mother, a high priestess in her candomblé community in Salvador, Bahia. Vagner's parents are Catholic; they pray to Mother Mary and light candles for their grandson. Vagner, who now wears a Saint Benedict medal[6] sent by his mother, says he 'tries to enjoy every day with Diego, as it may well be his last.'

Brazilian researchers analysed the association between pesticide use and congenital malformations between 1994 and 2014 in Cascavel and Francisco Beltrão, two cities close to the Falls with the highest pesticide use in Paraná. Results reveal that the incidence of congenital heart malformation was five times higher between 2004 and 2014 than in the preceding decade. The authors conclude that specific types of birth defects correlate with the increase in pesticide use (Dutra & Ferreira, 2017). Although male agricultural workers are most affected (Hendges et al., 2019), a study in the nearby state of Mato Grosso concludes that often both parents are exposed to pesticides; their children have a fourfold risk of birth defects such as spina bifida, leukaemia, and other forms of cancer (Ueker et al., 2016).[7]

From Small Farmer to Slum Dweller

Brazil possesses 12 per cent of the world's reserve of available freshwater. Chemical cartels mercilessly hush up news about human suffering from pesticide-contaminated ground water; there is ever so much (money/power) at stake. Research recently published by Public Eye[8] affirms that the most dangerous pesticides are used heavily in low- and middle-income countries (LMICs), despite being—for the most part—banned in Switzerland and the European Union (EU).

Public Eye's in-depth probe into the opaque world of highly hazardous pesticides also reveals that the Swiss agrochemical giant, Syngenta, is one of those mainly responsible for the flood of such products into LMICs. This conclusion is based on our analysis of exclusive industry data. . . . Brazil, the world's largest user, [exposes] millions to pesticides that present significant hazards to human health—including through [contaminated]

---

[6] According to Catholic tradition, the Saint Benedict medal is a symbolic protection against poisoning.

[7] Research elsewhere cites a clear correlation between pesticide exposure and poor neurodevelopment among children (van Wendel de Joode et al., 2016).

[8] Public Eye (formerly Berne Declaration) is a non-profit, independent Swiss watchdog organisation with around 25,000 members. Public Eye has been campaigning for more equitable relations between Switzerland and underprivileged countries for fifty years. Among its most important concerns are the global safeguarding of human rights, the socially and ecologically responsible conduct of business enterprises, and the promotion of fair economic relations.

drinking water. Scientists fear this could trigger an epidemic of chronic diseases. The time has come to put an end to this dirty business. (Gaberell & Hoinkes, 2019, p. 3)

The damage caused by agroindustrial practices goes far beyond illness from agrochemicals. The excessive use of pesticides also contributes to deforestation and land degradation. Inequality of land distribution accelerates the trend in land grabs for industrial agriculture, provokes the expulsion of small farmers, rural workers and their families, and their forced migration to crime-ridden slums and shantytowns in urban areas. And, finally, this damage creates even larger patches of dead land and water pushing more and more people to a systemic edge. Out of a total population of 208 million, almost nine million Brazilians are displaced (Forum for the Future, 2019).

Ritualising in Fearful Times: Harmful Pesticides

Ritualising concern over pesticide-induced illness, death, and injustice takes many forms in Brazil. The personal rituals performed by Marilea, Vagner, and their families help them feel like they are *doing something* to protect themselves and those they love (see Sympathetic state, Fig. 1). Public protest rituals use various art forms to effectively expose the truth about the misuse of dangerous pesticides. In 2017, a Brazilian samba school took on a political dimension with their performance alongside the lyrics of 'Xingu: A Cry from the Jungle': 'I am the forgotten son of the world. My heart is red in pain. I am the last immortal fighter—the true owner of this land. . .' This song defends the rights of indigenous people and decries the dangers of industrial farming, including abuse of pesticides (Prange, 2017).

Creative ritual protest may also take place in the nude (see Fig. 4). Brazilian artists use their naked bodies to warn not only of President Jair Bolsonaro's intolerance of feminism, homosexuality, and popular carnival songs like 'Xingu' but also of the environmental catastrophe ahead: 'In a polarised world, it feels like the naked truth. This resistance manifests itself in the naked body. In show after show, nudity takes on a political role' (Fisher, 2019). These art-filled rituals protest against the making of expulsions.

Revealing the truth about sensitive subjects is perilous. Investigative journalism, like that done by Public Eye, is dangerous for journalists, but it also puts at risk the lives of those they interview. Protestors took such risks as they marched against the excessive use of agrochemicals in Brasilia (see Fig. 5) and in front of Syngenta headquarters in Basel, Switzerland.

Today, when people are expelled from their professional livelihoods, living spaces, and the very biosphere that makes life possible, the situation cannot be fully understood in the usual terms of poverty and injustice, affirms Saskia Sassen. This is about 'new specialized geographies that cut across the old divides of North and South, East and West' (2015, p. 178).

[The spaces of the expelled] are marked by increasingly diverse groups, places, projects, and histories. In the past, the British Empire wanted the whole of Africa, and Spain wanted the

Fig. 4 Nudity takes on a political role. This photo reflects the spirit of the nude protests performed by Brazilian artists. | © Sasha Kargaltsev CC BY

Fig. 5 Protest march against excessive use of agrochemicals. Social movements and environmental organisations march in Brasilia to protest against changes in national laws that would permit an increase in the use of agrochemicals. | © Marcello Casal Jr/ABr CC BY

whole of Latin America, and so on. Today's powers want only specific components, and once done, they exit. These are mobile geographies that leave behind land and sites destroyed by their use, which then, in their extreme condition, are in fact expelled from these geographies of privilege: expelled to the zone of dead land and dead water. (Sassen, 2015, p. 178)

A look at how these expulsions are generated reveals the extent to which the sheer complexity of the global economy effectively scrambles the lines of responsibility for the displacements, evictions, and eradications it produces. This same convolutedness makes it equally hard for anyone who benefits from a global system to feel any personal or corporate responsibility for the depredations.

Case Study 3: Lead Contamination in Switzerland

Geneva has the dubious honour of being the third most expensive city in the world. It caters to people who can pay more for rent than most locals earn in a month. Our building was a squat that we helped upgrade in 1995 to low-income housing status thanks to a utopian project. With like-minded people, we banded together as a cooperative with a social conscience to promote participative, sustainable, and ecological low-rent housing.

On 16 August 2017 at seven o'clock in the evening, my family and I returned from holiday to find an explosion of dust[9] in our apartment. Our floors, walls, windows, ceilings, and all of our belongings were thickly coated with a heavy whitish powder that burned our bare hands and feet. This cloying powder hung in the air, irritating our sinuses and disrupting our digestive tracts. The source was a poorly confined worksite in the flat below; masons had removed ceilings and applied grinders, sanders, and sandblasting equipment to lead-laden plaster and old paint on woodwork and radiators.

In the immediate aftermath of the accident, our faces turned white, our fingers and toes tingled with numbness, our blood pressures rose and our muscles twitched (like a mobile phone vibrating in a jeans pocket); our vision, memory, and concentration were also affected. The worst was yet to come.

As is too often the case in toxic accidents, human suffering was aggravated by a cascade of poor decisions regarding renovation in an inhabited building, ignorance about human exposure, undetected contamination, and slow reactions by the authorities. It took six weeks for official confirmation that the ubiquitous powder was heavily laced with lead (Pb).[10] The Sisyphean task of ridding the deadly contaminant from the air and every surface and object in our home was aggravated by handicapping fatigue, persistent refusal of the owner and the masonry company to accept any responsibility for the accident, and acquiescent authorities.

---

[9] Under certain conditions, lead (Pb) may form combustible dust concentrations in the air (Atomized Products Group, 2013).

[10] After exposure to lead, the body rapidly tucks this dangerous element in the bone where it is, at least temporarily, immobilised and can do the least harm to vital organs (brain, kidneys, liver, heart). Lead levels in red blood cells (erythrocytes, not blood serum) should be measured within ten days of an acute exposure (within three days for children under three years of age). After that, minute traces may be found in the hair. Accurate and relatively safe methods for measuring bone lead exist today, but few countries have approved them for use *in vivo*.

From Utopia to Dystopia [11]

Over the years, the leadership of this flagship cooperative, known for promoting quality social, participative and ecological housing, imperceptibly adopted a corporate style of governance. Once the workday culture and rituals of the organisation became established, taken for granted, and finally applauded by local authorities, a series of unilateral, calamitous decisions revealed the widening distance of the leadership from its base and foundational ideals. [12] 'Risk management' effectively replaced the cooperative solutions that might have been born out of apology and sitting together around a table to determine how to prevent further contamination and suffering. When we asked about reimbursement for our cleaning and medical testing expenses, the response was immediate: 'Sue us'.

What happened to us is not an isolated event (see Fig. 6). [13] Lead paint was sold in Switzerland until 2005. Most buildings constructed or renovated before then have lead-painted surfaces. Impunity prevails over accountability. Our friends and family are concerned we'll face expulsion for speaking out. There is ever so much at stake. If Swiss authorities required owners to use the same precautions for work on lead-laden surfaces as it does, finally, for those contaminated with asbestos, renovation costs would soar. Yet, if we cannot find safety in our homes, where might we feel safe?

The dangers of lead contamination—identified already in Roman times—represent the tip of the iceberg. Very little is known about the chronic toxicity of the new, and mostly untested, building materials. The political, economic, scientific, and social implications of the environmental toxins produced today may well haunt our grandchildren and their children.

What if, together with our cooperative, our story was told to the press to encourage caution and build awareness of how easy it is to pollute living spaces? What if we devised creative solutions for such renovations? What if Switzerland made, and enforced, laws that treat toxic contaminates with the seriousness they merit? What if the Geneva toxicology department and hospital became leaders in innovative prevention and clean-up of the environmental toxins that contaminate our households? What if toxic contamination could be quickly measured and cleaned

---

[11] Dystopia [dystɔpi] (pl. dystopias) 1. a political or social concept referring to a utopia that turns into a disaster; a counter-utopia or anti-utopia. 2. A failed utopian achievement that causes intense suffering among the inhabitants who are subjected to the dysfunctional system. Synonym: disillusionment, disenchantment.

[12] See Knottnerus et al.'s analysis of the organisational process that led to the demise of Enron (2006). In this case, the issue is also mismanagement—not of finances per se but of the cooperative's foundational legacy of social ideals.

[13] The struggle for justice regarding toxic contamination of our homes may well be long. Nearly 50 years ago, asbestos was declared an undisputed cause of cancer. In 1989, the mineral was finally banned in Switzerland; the Swiss Schmidheiny family's Eternit Group commercialised asbestos products until 1994. Yet it still falls to Swiss people with asbestosis to prove they were contaminated. According to a European Union study, asbestos will have caused around half a million deaths in Europe by 2030 (Mariani, 2012).

Fig. 6 There is no safe level of lead exposure. The manufacture and sale of lead paint is still permitted in over 60% of the countries across the world. Inhalation of the dust generated by the preparation of lead-painted surfaces for repainting contaminates inhabitants, workers, and their children. The focus of the 2020 World Health Organization's week of action is a global phase-out of lead paint through regulatory and legal measures. | © World Health Organisation (WHO) Public Domain

up? What if victims were rapidly tested, correctly diagnosed, and nursed back to physical and psychological health?

Ritualising in Fearful Times: Our Homes Under Threat

At first, my family and I used dark humour to relieve our fears and our tears of frustration. Endless cleaning, unsatisfying legal procedures with loaded die, and consultations with doctors no experience in treating Saturnism did little to restore our sense of safety, or to (re)incorporate us in the cooperative. Although we had no ready-made rituals at our disposal, we felt a compelling need to *do something more* to push back our fears, recurring nightmares, and anxiety, to reclaim our space, our physical and emotional health, and to restore broken connections with our neighbours.

The ordinary spaces in our home began to feel extraordinary as the thick dusty air was treated with humidifiers (to bring down the particles of lead powder), then cleansed by industrial-strength air filters—and perfumed with essential oils. Floor-boards, a main vector for the lead powder, were covered over with plastic—and sprinkled with purified water and salt, symbols of our tears. Another vector of contamination, the light shaft at the end of our hallway, needed our attention too. In counterpoise to the invasive plague, words of life and hope now emanate from the windows: serenity, joy, purity, light, being heard, harmony, beauty, respect, life, justice, and reparation. Firefly lights chase the deadly shadow of death; each week the space is beautified with fresh flowers (see Fig. 7).

My husband joined the committee that manages our building. Once we feel stronger, we intend to invite our friends and neighbours for a Sunday-in-pyjamas brunch and are planning a creative public protest for the International Lead Poison-ing Prevention Week. A story-strip of our misadventure will be distributed widely at the brunch and the protest.

A colleague pointed out that this chapter too is a form of ritualising our fear and helplessness: 'By publishing this text you are completing a defensive response that was interrupted by illness, general indifference to the contamination, and unsatisfying legal proceedings.' This response does indeed contribute to restoring my sense of dignity and safety. [14]

Regrettably, we are not the only ones who feel unsafe in our home.

No Place Left to Go

Some time ago, Anne, a friend who lives near Amsterdam in a beautiful house with a lovely garden, began coughing; her condition slowly worsened. The source of her cough has been tied to airport pollution; their home is under the flight paths of aircraft flying into Schiphol, one of the busiest airports in Europe. [15] Reluctantly, Anne searched for a house in another area, only to find that a new extension of the airport would put them in the same situation. Conscientiously, Anne restricts her travel to ground transport. Her need to *do something more* has led her to curate art exhibitions on subjects related to the dilemma she faces (consumerism, death).

---

[14] Unlike the other situations described here, I can explain the effect on wellbeing of rituals practised with embodied intentionality because I have felt it myself. It took more than two years to write this chapter. With the finishing touches, I did indeed feel a physical release or discharge.

[15] Nine out of ten people worldwide breathe polluted air (World Health Organization, 2018).

Fig. 7 Ritualising hopes and fears. Ever since our home was contaminated with high levels of lead, fresh flowers beautify a light shaft, one of the vectors of infiltration. The words pasted on these windows bear witness to our desire for serenity, joy, harmony, light, purity, and respect, as well as for reparation and justice. Above all, we want to feel safe in our own home. | © J. Gordon-Lennox

Singing in a choir composed of asthmatics is yet another way Anne meets her needs for self- and co-regulation.

In the wake of the destructive fires that skirted her home in Sonoma, California, Carrie Kramlich contacted me for advice on creating rituals to cope with post-disaster trauma. Later, she wrote:

> Northern California is like a resurrected Pompeii, right now. The fired turned us all to stone. The heat and scourge of the lava cooled and as it did, people started moving slowly... then waking up. And what's coming out of it is spectacular. My mother passed away last spring. We had a house full of furniture and tons of things to give away. When the fires came in early October we were still grieving. Just days after the fires stopped burning my sister and I let all of my mom's things fly out of her house to dozens of wildfire survivors. We were like tornados in reverse motion. You should come see. (Carrie Kramlich, personal communication, 2017)

Carrie is convinced that only truly secular rituals can offer diverse populations of survivors a sense of relief and safety. 'That's not possible with religious rituals because they tend to stir up irrelevant questions like: Did we deserve this? Why is god punishing me?' (Carrie Kramlich, personal communication, 2019).

Ritualising at the Systemic Edge

The situation of people pushed to a systemic edge signals subterranean trends that are related not only to politics and economics but also to the spiritual dimensions of contemporary human experience. The challenges of ritualising in fearful times, particularly during ongoing threat, are not the same for those who belong to mainstream spiritual systems, for the socially excluded, and for the expelled.

Small cash-crop farmers in India, pushed to the limit by an unsustainable agricultural model, who die by suicide before declaring bankruptcy, are completely invisible to the system (Sassen, 2011). Such deaths represent sharp edges that point to a huge but subtle trend: the hazards with which farmers are forced to live represent a global disaster that crosses all cultures. [16] Tracing responsibility for these farmers' circumstances, as well as for the situations in the cases described above, is difficult, if not impossible. Those who benefit rarely feel answerable to the victims; seldom is anyone held accountable.

Even people firmly anchored in mainstream systems may find traditional rituals ineffectual as their situation deteriorates, or when several systemic edges converge. While the Seti Khola disaster did not fundamentally change Sumila and her family's situation, it pushed Sarila into the precarious state of widowhood and social exclusion. Diego's condition does not seriously affect his father's or grandparents' daily routines,

---

[16] Suicide born of the despair brought about by the farmer's individual, or even collective, expulsion must be differentiated from ritual suicide. Suicide attacks or self-immolation are performed as radical political or religious protest, martyrdom, or sacrifice 'on behalf of a collective cause' (Biggs, 2006, p. 173).

but it isolates his mother socially and excludes her from the workforce. On the spiritual level, most of these people still count on some mainstream or parallel spiritual system. Yet the closer people move to a systemic edge, the higher the likelihood this spiritual support will also fail them. As the expelled become invisible statistically, geographically, politically, and economically, they disappear from their spiritual systems too.

Global changes in spatial, temporal, environmental, climatic, and spiritual contexts affect our resources for coping with threat and ongoing danger. The amazing capacity of the human imagination to produce and embody abstract signs is crucial both to the creative processes of ritualmaking and to healing trauma. Porges' research on polyvagal theory (2011) gives us new insights into the *how* and *why* of ritual practice. As we 'make special' we respond to our human physical, psychological, and spiritual need to overcome our ambivalence about closeness and to establish trusting social relationships that ensure our survival.

Conclusion

Our world is increasingly unpredictable. Human-made disasters occur regularly on large and small scales around the world. Opportunities to feel safe and secure are fundamental to human wellbeing. Being safe and feeling safe are not the same thing. We are at risk for psychosocial trauma when our sense of control over our own physical environment (home, workplace, body) is compromised by people or events that threaten to seriously harm or even kill us or those we love.

The cases presented in this chapter demonstrate how a disaster can push ordinary people out of the common to a systemic edge. Before the disaster, the central figure(s) may not have enjoyed full incorporation in their respective communities, yet any exclusion they experienced took place within a micro or macro system. Their role as victim in a disaster they did not create and cannot control may incorporate and anchor them more firmly in their community. Too often disaster moves them to an extreme condition at a systemic edge where they are in even greater danger—that of being expelled from the system completely and, even worse, of feeling less than human, of becoming invisible.

Human beings have always used ritual to face the transience of life. Today, the ephemerality of life is too often due to human-made disasters. Yet individuals, collective groups, and even global systems still construct and do ritual to create islands of safety within and among us. Ritualising is an underexplored resource for coping with chronic illness and ongoing disasters, for healing the ensuing trauma and for restoring broken connections.

References

Atomized Products Group. (2013). *Lead powder safety data sheet.* Accessed August 30, 2020, from www.atomizedproductsgroup.com/wp-content/uploads/2015/03/Lead_Powder_SDS_NA_MEX_112113_FINAL.pdf

Biggs, M. (2006). Dying without killing: Self-immolations, 1963–2002. In D. Gambetta (Ed.), *Making sense of suicide missions.* Oxford University Press.

Bhandari, R. B. (2014). Social capital in disaster risk management: A case study of social capital mobilization following the 1934 Kathmandu Valley earthquake in Nepal. *Disaster Prevention and Management, 23*(4), 314–328.

Bhandari, R. B., Okada, N., Yokomatsu, M., & Ikeo, H. (2010). Interpreting an urban ritual event in terms of improving the capacity to cope with disaster risk: A case study of Kathmandu. *Journal of Natural Disaster Science, 32*(1), 31–42.

Bhandari, R. B., Okada, N., & Knottnerus, J. D. (2011). Urban ritual events and coping with disaster risk a case study of Lalitpur, Nepal. *Journal of Applied Social Science, 5*(2), 13–32.

Dhimal, M. L. (2015). *Gender differentiated health impacts of environmental and climate change in Nepal* (pp. 96–100). International Conference on Climate Innovation and Resilience for Sustainable Livelihood. Kathmandu, .

Dutra, L. S., & Ferreira, A. P. (2017). Associação entre malformações congênitas e a utilização de agrotóxicos em monoculturas no Paraná, Brasil. *Saúde Debate, 41*(n. especial), 241–253. Accessed August 17, 2020, from www.scielo.br/pdf/sdeb/v41nspe2/0103-1104-sdeb-41-spe2-0241.pdf

Evetts, C. (2017). Fight or flight versus tend and befriend behavioral response to stress. *American Journal of Occupational Therapy, 71.*

Fisher, M. (2019, March 29). The shock of the nude: Brazil's stark new form of political protest. *The Guardian.* Accessed August 17, 2020, from www.theguardian.com/stage/2019/mar/29/shock-of-the-nude-brazil-festival-protest-theatre

Forum for the Future. (2019). *The future of sustainability 2019.* Website. Accessed August 17, 2020, from https://thefuturescentre.org/fos2019/#

Gaberell, L., & Hoinkes, C. (2019). *Highly hazardous profits: How Syngenta makes billions by selling toxic pesticides.* Public Eye Report. Accessed August 17, 2020, from www.publiceye.ch/en/media-corner/press-releases/detail/syngenta-makes-billions-selling-hhps

Hendges, C., Schiller, A. D. P., Manfrin, J., Macedo, E. K., Jr., Gonçalves, A. C., Jr., & Stangarlin, J. R. (2019). Human intoxication by agrochemicals in the region of South Brazil between 1999 and 2014, Journal of Environmental Science and Health, Part B, *54*(4), 219–225.

Kargel, J. (2014). *One scientist's search for the causes of the deadly Seti River flash flood.* NASA Earth Observatory Blog, Notes from the Field. Accessed August 17, 2020, from https://earthobservatory.nasa.gov/blogs/fromthefield/2014/01/24/setiriverclues

Knottnerus, J. D., Ulsperger, J. S., Cummins, S., & Osteen, E. (2006). Exposing Enron: Media representations of ritualized deviance in corporate culture. *Crime, Media, Culture, 2*(2), 177–195.

Malinowski, B. (1948 [1925]). *Magic, science, and religion.* Free Press.

Mariani, D. (2012). *Dying from asbestos. . . and having to prove it.* Swissinfo.ch. Accessed August 17, 2020, from www.swissinfo.ch/eng/dying-from-asbestos%2D%2Dand-having-to-prove-it/32061028

Pierre-Louis, K. (2017). There's actually no such thing as a natural disaster. Hazards are natural; disasters are manmade. *Popular Science.* Accessed August 17, 2020, from www.popsci.com/no-such-thing-as-natural-disaster

Porges, S. W. (2011). *The polyvagal theory: Neurophysiological foundations of emotions, attachment, communication, and self-regulation.* WW Norton & Company.

Porges, S. W. (2015). Making the world safe for our children: Down-regulating defence and up-regulating social engagement to 'optimise' the human experience. *Children Australia, 40*(2), 114–123.

Prange, A. (2017). *Rio: Carnival dancers with a message*. DW Made for Minds. Accessed August 17, 2020, from https://p.dw.com/p/2YLzq

Sassen, S. (2011). The global street: Making the political. *Globalizations, 8*(5), 573–579.

Sassen, S. (2014). *Expulsions: Brutality and complexity in the global economy*. Harvard University Press.

Sassen, S. (2015). At the systemic edge. *Cultural Dynamics, 27*(1), 173–181.

South Asia Network on Dams, Rivers and People (SANDRP). (2014, January 26). Explained: Seti River floods in May 2012, Nepal: A chain of events, starting at 25,000 feet! Accessed August 17, 2020, from https://sandrp.in/2014/01/26/explained-seti-river-floods-in-may-2012-nepal-a-chain-of-events-starting-at-25000-feet

Time Magazine. (1957). *Psychology of brainwashing*. Accessed August 17, 2020, from http://content.time.com/time/magazine/article/0,9171,824959,00.html.

van Wendel de Joode, B., Mora, A. M., Lindh, C. H., . . . Mergler, D. (2016). Pesticide exposure and neurodevelopment in children aged 6–9 years from Talamanca, Costa Rica. *Cortex, 85*, 137–150.

Ueker, M. E., Silva, V. M., Moi, G. P., Pignati, W. A., Mattos, I. E., & Silva, A. (2016). Parenteral exposure to pesticides and occurrence of congenital malformations: Hospital-based case-control study. *BMC Pediatrics, 16*(1), 125.

World Health Organization (WHO). (2018, May 2). *9 out of 10 people worldwide breathe polluted air, but more countries are taking action*. Accessed August 17, from www.who.int/news-room/detail/02-05-2018-9-out-of-10-people-worldwide-breathe-polluted-air-but-more-countries-are-taking-action

Jeltje Gordon-Lennox, MDiv, is a psychotherapist trained in body-based approaches and world religions. Her research and practice are influenced by her life experiences in conflict zone on several continents, in particular her work with the International Committee of the Red Cross. She has written five practical guides on secular ritualising, two in French and three in English. This collection continues the conversation on ritual and trauma begun in *Emerging Ritual in Secular Societies: A Transdisciplinary Conversation* (2017, Jessica Kingsley Publishers). Jeltje lives with her husband and their two children in Switzerland. *Website:* gordon-lennox.ch *E-mail:* jeltje@gordon-lennox.ch

Index[1]

A

acoustic stimulation
    low-frequency sounds, 60
    neuroception of safety, 59–60
    safety signals and calmness, 59–60
Action Plan for Peace (Afghanistan), 128
Adamovich, Ales, 178
addiction, 151–160
    as adaptation to dislocation, 155–157
    cause as societal malfunction, 153–155
    definition, 152
    denial *vs.* acceptance, 158
    drugs, 160
    examples of, 16–17, 151–152
    and global capitalism, 191
    historical perspective, 152–153
    in modern economy, 154–157
    in the modern era, 153
    to power, 17
    recovery, 157–158
    social change, 158–160
    societal fragmentation, 153–155
    stress diseases, 156
Afghanistan Centre for Memory, 133*n*5
Afghanistan conflict, 125–134
    conflict mapping project, 128
    flags, 133
    legislation (Amnesty Law), 128
    memory boxes, 16, 129–133
    monuments and museums, 126, 127
    peace activism, 163
    post-conflict context, 134
    power struggles, 126
    transitional justice failings, 125, 128–129
    victims' museum, 127
    war criminals, 128
Afghanistan Human Rights and Democracy
    Organization (AHRDO), 132, 133
Afghanistan Independent Human Rights
    Commission (1), 125*n*1, 127, 128
African-Americans
    church shooting (white supremacy), 165
    murder victim (by police), 171, *172–173*
    pandemic cases and inequality, 2
    violent extremism response (US presidential
        election), 165–167
agricultural food production and radioactive
    fallout, 176, 178
agrochemical companies, 198–200
airport pollution, 205, 205*n*15
Al Qaeda, 163
Alexander, Bruce, 16, 87, 88
alginates (seaweed), 183, 186
Alonso, A.M., 126
Amnesty Law (Afghanistan), 128
ancient rituals, 15
animal species
    and grief, 27
    social ritualisation, 27–28
    trauma, 192, 194*n*3
Annan, Kofi, 183
apes and social ritualisation, 27

---

[1] Page numbers in italics refer to figures; The indication mark (*n*) refer to footnote.

Ingram Content Group UK Ltd.
Milton Keynes UK
UKHW022025300323
419441UK00001B/9